Antibiotic Resistance in *Acinetobacter* and Associated Treatment Strategies

Antibiotic Resistance in *Acinetobacter* and Associated Treatment Strategies

Guest Editors

Steven E. Fiester
William Penwell

MDPI

Basel • Beijing • Wuhan • Barcelona • Belgrade • Novi Sad • Cluj • Manchester

Guest Editors
Steven E. Fiester
Department of Biological Sciences
Florida Gulf Coast University
Fort Myers, FL
USA

William Penwell
Department of Biology and Marine Science
Jacksonville University
Jacksonville, FL
USA

Editorial Office
MDPI AG
Grosspeteranlage 5
4052 Basel, Switzerland

This is a reprint of the Special Issue, published open access by the journal *Antibiotics* (ISSN 2079-6382), freely accessible at: https://www.mdpi.com/journal/antibiotics/special_issues/Y5TR37D7EV.

For citation purposes, cite each article independently as indicated on the article page online and as indicated below:

Lastname, A.A.; Lastname, B.B. Article Title. *Journal Name* **Year**, *Volume Number*, Page Range.

ISBN 978-3-7258-5027-3 (Hbk)
ISBN 978-3-7258-5028-0 (PDF)
https://doi.org/10.3390/books978-3-7258-5028-0

Contents

antibiotics

MDPI

Editorial

Combating the Growing Threat of *Acinetobacter baumannii* Resistance

Eduarda Kffuri Goncalves [1], William Frank Penwell [2] and Steven Eugene Fiester [1,3,4,5,6,7,*]

1 Department of Biological Sciences, Florida Gulf Coast University, Fort Myers, FL 33965, USA; ekkffurigoncalve6643@eagle.fgcu.edu
2 Department of Biology and Marine Science, Jacksonville University, Jacksonville, FL 32211, USA; wpenwel@ju.edu
3 Family Medicine Residency Program at Lee Health, Florida State University College of Medicine, Fort Myers, FL 33901, USA
4 Internal Medicine Residency Program at Lee Health, Florida State University College of Medicine, Cape Coral, FL 33991, USA
5 Department of Chemistry, Furman University, Greenville, SC 29617, USA
6 School of Health Research, Clemson University, Clemson, SC 29634, USA
7 Department of Pathology, Prisma Health Upstate, Greenville, SC 29601, USA
* Correspondence: sfiester@fgcu.edu

1. Introduction

Antimicrobial resistance remains one of the most urgent and complex challenges facing global public health in the 21st century. Among antibiotic resistant bacteria, *Acinetobacter baumannii* is considered a particularly alarming pathogen due to its propensity for drug resistance, making this bacterium a significant threat to patient safety and hospital infection control efforts [1]. Due to its resistance capability, *A. baumannii* is classified as one the ESKAPE pathogens, a group of six highly drug-resistant bacterial pathogens notorious for their ability to evade the effects of antimicrobial agents as well as the host immune response [2].

A. baumannii itself is a Gram-negative, aerobic bacterium commonly isolated in the clinical environment and generally associated with ventilator-associated pneumonia, bloodstream infections, urinary tract infections, meningitis, and wound infections in critically ill or immunocompromised patients [3]. The success of *A. baumannii* as a pathogen is, however, not solely due to its resistance phenotypes but also due to its ability to evade immune responses, tolerate desiccation, resist disinfectants, and form biofilms on surfaces and medical devices [4]. Taken together, these phenotypes compounded by this pathogen's genetic plasticity make *A. baumannii* uniquely suited to thrive in healthcare environments. The global rise in multidrug-resistant (MDR), extensively drug-resistant (XDR), and pandrug-resistant (PDR) *A. baumannii* strains highlights an urgent need for innovative treatment options, deeper insights into resistance mechanisms, and strengthened international surveillance initiatives.

In recognition of the growing threat of drug-resistant *A. baumannii*, the World Health Organization has designated carbapenem-resistant *A. baumannii* (CRAB) as a critical priority pathogen for research and drug development due to its increasing prevalence and severe impact in healthcare settings worldwide [5]. This classification underscores the necessity for a coordinated response to mitigate this pathogen's spread and improve treatment outcomes. *A. baumannii* has demonstrated an exceptional capacity to resist antimicrobial treatments. Its arsenal of resistance determinants includes the production of β-lactamases (particularly oxacillinases), the use of multidrug efflux pumps, alterations in outer membrane permeability, and the capacity to form biofilms. These resistance mechanisms do not act in isolation; rather, they interact synergistically, conferring high levels of resistance to diverse antibiotics, including last-resort agents like colistin [6].

The eight contributions included in this Special Issue collectively highlight the multifaceted approaches required to address *A. baumannii*, emphasizing the value of genomic surveillance, stringent infection control practices, strategic antimicrobial stewardship, and standardized testing for disinfectant resistance. Collectively, these contributions explore key dimensions of the *A. baumannii* challenge, including its antibiotic resistance mechanisms, the role of disinfectant protocols in outbreak prevention, the emergence of tolerance to biocidal agents, the influence of global clones on resistance phenotypes, and the assessment of current and experimental therapies, both as monotherapies and in combination regimens.

2. Overview of Published Papers

Genomic surveillance has proven essential for understanding the distribution and evolution of antimicrobial resistance. The article by Hamed et al. (contribution 1) conducted a comprehensive genomic study in Egypt, which revealed a distribution of global clones (GCs) among clinical isolates of *A. baumannii*, with global clone 2 (GC2) emerging as the dominant lineage in the resistome landscape. Isolates within this lineage demonstrated consistent resistance to multiple antibiotics, including amikacin, colistin, and trimethoprim. Further genomic analysis identified 51 resistance determinants across 10 antimicrobial classes. Genes such as $bla_{\text{OXA-23}}$, $bla_{\text{NDM-1}}$, and *armA* were widespread, while mutations in *carO* correlated strongly with carbapenem resistance. Among the resistance-associated genes, $bla_{\text{OXA-23}}$ was the most prevalent and a key driver of resistance in clinical isolates. These findings underscore the importance of whole-genome sequencing in mapping resistance patterns and tailoring region-specific treatment and prevention strategies.

Beyond intrinsic genetic resistance, *A. baumannii* exhibits adaptive response mechanisms to environmental stressors, a feature that further complicates clinical management. One such mechanism, known as adaptive efflux-mediated resistance, enables the bacterium to upregulate efflux pump systems in response to biological signals, such as exposure to human serum. In the article by Young et al. (contribution 2), the authors demonstrate that treatment of phenotypically susceptible strains in vivo can result in failure through this mechanism. This research identified a regulatory gene encoding a putative YhaK protein as essential to this process. The mutation of *yhaK* impaired efflux activity and increased intracellular drug accumulation, confirming its regulatory function. Notably, *yhaK* expression was significantly upregulated in human serum compared to standard media, highlighting its role in environmental adaptation and specific resistance to fluoroquinolones and tetracyclines in the human host.

The adaptability of *A. baumannii* extends further to traits that support biofilm formation, a key factor in its persistence on hospital surfaces and devices. Abd El-Rahman et al. (contribution 3) evaluated the contribution of efflux pump genes, particularly those belonging to the resistance-nodulation-division (RND) family, in biofilm development. While this study did not demonstrate a direct influence of these genes on biofilm formation, it pointed to the potential of further exploring efflux systems as therapeutic targets. Further understanding the role of efflux pumps in the formation of biofilms may offer future benefits in combating antimicrobial resistance and reducing environmental persistence.

In addition to exposure to antibiotics, clinical isolates of *A. baumannii* may also be exposed to various disinfectants and have developed resistance to many of these as well. Kelemen et al. (contribution 4) investigated the molecular basis of disinfectant resistance during an outbreak of CRAB in an intensive care unit in Hungary. This study demonstrated that modification of the disinfection protocol, specifically by replacing a quaternary ammonium compound (QAC)-based disinfectant (alkyl dimethyl benzyl ammonium chloride (C12-16 ADBAC), also known as MBF) with an alcohol-based disinfectant (10–20% 2-phenoxy-ethanol and 5–10% benzalkonium chloride, also known as IP), played a pivotal role in controlling the outbreak. In this study, six CRAB isolates from six individual patients all exhibited similar resistance profiles, showing resistance to carbapenems, aminoglycosides, and fluoroquinolones, while remaining susceptible to colistin. Genomic analysis of these isolates also revealed the presence of *qacE* and *qacE*Δ*1*, which encode transmembrane proteins responsible for QAC disinfectant tolerance. The products of these genes also contribute to broader biocide resistance by promoting efflux pump activity. The switch to an alcohol-based disinfectant, combined with stricter containment protocols, successfully ended the outbreak with no new infections reported, contrasting with more prolonged outbreaks when previous QAC-based protocols were utilized. This study underscores the importance of recognizing genetic mechanisms of disinfectant tolerance and suggests incorporating disinfectant susceptibility testing into revised infection control and antimicrobial stewardship programs.

Despite its historical use as a last-resort therapy due to nephrotoxicity, colistin has regained relevance for treating *A. baumannii* infections due to this pathogen's resistance to nearly all other antibiotic classes, including carbapenems. The limited penetration of intravenous (IV) colistin into pulmonary tissues has, however, prompted exploration of alternative delivery methods. The study by Andrianopoulos et al. (contribution 5) investigated the efficacy of IV colistin in combination with high-dose nebulized colistin in treating ventilator-associated pneumonia secondary to *A. baumannii* bacteremia. Patients receiving this combination therapy showed significantly lower 7-day and 28-day mortality rates, greater sepsis resolution, and earlier vasopressor discontinuation compared to those treated with IV colistin alone. These findings reinforce the potential benefits of combination therapies as well as more targeted and localized antibiotic delivery methods to overcome tissue penetration barriers and combat MDR infections.

Lim et al. (contribution 6) investigated the potential use of fosfomycin in combination with sulbactam or colistin against clinical isolates of *A. baumannii* resistant to both antibiotics individually. While fosfomycin alone lacks reliable bactericidal activity against *A. baumannii*, its ability to inhibit early stages of cell wall synthesis makes it a valuable adjunct in combination regimens. This study demonstrates significant in vitro synergy when fosfomycin was combined with sulbactam, particularly in isolates that had previously shown resistance to sulbactam monotherapy. This synergistic effect is believed to arise from fosfomycin's disruption of peptidoglycan biosynthesis, which may enhance sulbactam's ability to bind to penicillin-binding proteins. Additionally, this combination therapy was noted to reduce the minimum inhibitory concentrations of both drugs, potentially restoring clinical efficacy. These findings are critical in the context of limited treatment options for drug-resistant *A. baumannii*, underscoring the importance of re-evaluating older antibiotics like fosfomycin within strategic combination therapies to overcome complex resistance phenotypes and improve patient outcomes.

A detailed review of current therapeutic approaches for CRAB infections, particularly in light of emerging β-lactam agents, is provided by Rafailidis et al. (contribution 7). This review emphasizes the limitations of traditional antibiotics such as colistin and tigecycline when used as monotherapies. Colistin, though frequently deployed as a last-resort therapeutic, poses significant challenges due to its nephrotoxicity and limited penetration into pulmonary tissues when delivered intravenously. Tigecycline, while offering utility in the treatment of soft tissue infections, is less effective in the treatment of bacteremia because it does not achieve high serum concentrations. The authors argue for the increased efficacy of combination regimens, especially those incorporating agents with complementary mechanisms of action. Notably, they highlight the potential benefits of treating with sulbactam–durlobactam, a novel β-lactam/β-lactamase inhibitor pairing, which demonstrates intrinsic activity against *A. baumannii* and overcomes class D β-lactamase-mediated resistance. Another promising agent discussed is cefiderocol, which exhibits high stability against all four classes of β-lactamases and uses a siderophore-mediated entry system to reach its target.

Building on the aforementioned considerations, the review by Serapide et al. (contribution 8) examines the pharmacokinetic and pharmacodynamic (PK/PD) challenges associated with treating CRAB, particularly in critically ill patients. This review demonstrates how suboptimal tissue concentrations of antibiotics, even when used to treat phenotypically susceptible isolates, can lead to therapeutic failure. This review reinforces the necessity of optimizing dosing strategies, especially for agents like cefepime and tigecycline, to ensure effective drug levels at infection sites. This review also underscores the clinical potential of cefiderocol, citing its effective penetration and unique mechanism of action wherein bacterial iron uptake pathways are exploited to gain access to the periplasmic space. In addition to its broad stability against β-lactamases, cefiderocol is associated with lower mortality rates when compared to colistin-based therapies. This review reemphasizes concerns over emerging resistance and the importance of antibiotic stewardship to preserve the efficacy of limited therapeutic interventions.

3. Discussion and Conclusions

This Special Issue underscores the complexity and clinical urgency of combating *A. baumannii*, particularly CRAB strains. The featured contributions highlight a multifaceted public health challenge, where this bacterium's ability to persist in healthcare settings, resist both antibiotics and disinfectants, and rapidly adapt to environmental pressures, allows this pathogen to continually overcome current countermeasures [1,5]. Collectively, these studies reinforce the importance of genomic surveillance in identifying resistance mechanisms and tracking the spread of high-risk *A. baumannii* lineages to inform region-specific treatment protocols [7]. Insights into adaptive resistance, such as serum-induced efflux mechanisms and increased environmental tolerance, highlight the fact that reduced antibiotic clinical efficacy is driven not only by classical antibiotic resistance mechanisms, but also by dynamic gene regulation and stress response pathways. The ability of *A. baumannii* to adapt to various environments is strengthened by the accumulation of resistance determinants through both horizontal and vertical gene transfer, enabling rapid dissemination of resistance within and across bacterial populations. Genomic characterization of *A. baumannii* is therefore essential to decipher the genetic makeup of resistant strains and select more effective, isolate-specific treatment strategies [8]. Ultimately, infection control strategies that consider this pathogen's resistance to both antibiotics and disinfectants, whose resistance mechanisms could be shared, are critically needed.

While disinfection protocols are essential for curbing the spread of *A. baumannii*, mitigation of this pathogen's ability to establish an infection and the achievement of improved clinical outcomes demand a more comprehensive understanding of the organism's pathogenicity and resistance mechanisms. The success of *A. baumannii* as a nosocomial pathogen lies in its multifaceted virulence strategies, which include the production of β-lactamases, alterations in outer membrane proteins, activation of multidrug efflux pumps, and robust biofilm formation. These mechanisms work synergistically to enhance the organism's survival in hostile environments, contributing to its persistence and complicating eradication efforts [9]. These challenges underscore the necessity for continuous evaluation of current therapeutic strategies, deeper investigation into the potential of combination therapies, and the development of novel antimicrobial agents.

In response to difficult-to-treat resistant (DTR) and PDR *A. baumannii*, combination therapy has emerged as a cornerstone strategy. Empirical regimens often begin with high doses of ampicillin–sulbactam or tigecycline. Polymyxins such as colistin are also commonly employed, despite their known limitations, including nephrotoxicity and poor pulmonary penetration. Newer agents like sulbactam–durlobactam (SUL/DUR) and cefiderocol have shown considerable promise in treating CRAB infections. SUL/DUR has demonstrated favorable activity and may be a preferable option for critically ill patients, particularly when other alternatives are limited [10]. In critically ill patients, optimizing PK/PD parameters is vital; for example, transitioning from intermittent to continuous infusion of β-lactams has been associated with improved clinical outcomes [11]. These strategies highlight the importance of individualized therapy guided by antimicrobial susceptibility profiles and patient-specific factors, reinforcing the critical role of antimicrobial stewardship programs.

In summary, this Special Issue offers a comprehensive overview of the attributes and challenges presented by A. baumannii. These studies advance our understanding of available treatments and shed light on mechanisms that drive immune evasion and lead to antibiotic failure. Emphases on early detection and surveillance strategies for both antibiotic and disinfectant resistance highlight emerging approaches in infection control. Ultimately, the contributions within this Special Issue aim to drive continued research into novel therapeutic development, optimization of current drug combinations, and identification of genetic determinants that drive resistance and persistence. Only through interdisciplinary collaboration between fields such as basic microbiology, pharmacology, and epidemiology can we outpace the evolving threat posed by this highly adaptive pathogen.

Conflicts of Interest: The authors declare no conflicts of interest.

List of Contributions

1. Hamed, S.M.; Elkhatib, W.F.; Brangsch, H.; Gesraha, A.S.; Moustafa, S.; Khater, D.F.; Pletz, M.W.; Sprague, L.D.; Neubauer, H.; Wareth, G. *Acinetobacter baumannii* global clone-specific resistomes explored in clinical isolates recovered from Egypt. *Antibiotics* **2023**, *12*, 1149.
2. Young, M.; Chojnacki, M.; Blanchard, C.; Cao, X.; Johnson, W.L.; Flaherty, D.; Dunman, P.M. Genetic determinants of *Acinetobacter baumannii* serum-associated adaptive efflux-mediated antibiotic resistance. *Antibiotics* **2023**, *12*, 1173.
3. Abd El-Rahman, O.A.; Rasslan, F.; Hassan, S.S.; Ashour, H.M.; Wasfi, R. The RND efflux pump gene expression in the biofilm formation of *Acinetobacter baumannii*. *Antibiotics* **2023**, *12*, 419.
4. Kelemen, J.; Sztermen, M.; Dakos, E.K.; Budai, J.; Katona, J.; Szekeressy, Z.; Sipos, L.; Papp, Z.; Stercz, B.; Dunai, Z.A. Complex Infection-Control Measures with Disinfectant Switch Help the Successful Early Control of Carbapenem-Resistant *Acinetobacter baumannii* Outbreak in Intensive Care Unit. *Antibiotics* **2024**, *13*, 869.
5. Andrianopoulos, I.; Kazakos, N.; Lagos, N.; Maniatopoulou, T.; Papathanasiou, A.; Papathanakos, G.; Koulenti, D.; Toli, E.; Gartzonika, K.; Koulouras, V. Co-Administration of High-Dose Nebulized Colistin for *Acinetobacter baumannii* Bacteremic Ventilator-Associated Pneumonia: Impact on Outcomes. *Antibiotics* **2024**, *13*, 169.
6. Mohd Sazlly Lim, S.; Heffernan, A.; Naicker, S.; Wallis, S.; Roberts, J.A.; Sime, F.B. Evaluation of fosfomycin-sulbactam combination therapy against carbapenem-resistant *Acinetobacter baumannii* isolates in a hollow-fibre infection model. *Antibiotics* **2022**, *11*, 1578.
7. Rafailidis, P.; Panagopoulos, P.; Koutserimpas, C.; Samonis, G. Current therapeutic approaches for multidrug-resistant and extensively drug-resistant *Acinetobacter baumannii* infections. *Antibiotics* **2024**, *13*, 261.
8. Serapide, F.; Guastalegname, M.; Gullì, S.P.; Lionello, R.; Bruni, A.; Garofalo, E.; Longhini, F.; Trecarichi, E.M.; Russo, A. Antibiotic treatment of Carbapenem-resistant *Acinetobacter baumannii*

infections in view of the newly developed β-lactams: a narrative review of the existing evidence. *Antibiotics* **2024**, *13*, 506.

References

1. CDC. *Antibiotic Resistance Threats in the United States, 2019*; Centers for Disease Control and Prevention Office of Infectious Disease: Atlanta, GA, USA, 2019.
2. Boucher, H.W.; Talbot, G.H.; Bradley, J.S.; Edwards, J.E.; Gilbert, D.; Rice, L.B.; Scheld, M.; Spellberg, B.; Bartlett, J. Bad bugs, no drugs: No ESKAPE! An update from the Infectious Diseases Society of America. *Clin. Infect. Dis.* **2009**, *48*, 1–12. [PubMed]
3. Howard, A.; O'Donoghue, M.; Feeney, A.; Sleator, R.D. *Acinetobacter baumannii*: An emerging opportunistic pathogen. *Virulence* **2012**, *3*, 243–250. [PubMed]
4. Nowak, J.; Paluchowska, P. *Acinetobacter baumannii*: Pathogenesis, antimicrobial resistance, and treatment options. *Adv. Clin. Exp. Med.* **2016**, *25*, 951–957.
5. WHO. *Global Priority List of Antibiotic-Resistant Bacteria to Guide Research, Discovery, and Development of New Antibiotics*; WHO: Geneva, Switzerland, 2017.
6. Vrancianu, C.O.; Gheorghe, I.; Czobor, I.B.; Chifiriuc, M.C. Antibiotic resistance profiles, molecular mechanisms and innovative treatment strategies of *Acinetobacter baumannii*. *Microorganisms* **2020**, *8*, 935. [CrossRef] [PubMed]
7. Higgins, P.G.; Dammhayn, C.; Hackel, M.; Seifert, H. Global spread of carbapenem-resistant *Acinetobacter baumannii*. *J. Antimicrob. Chemother.* **2010**, *65*, 233–238. [PubMed]
8. Wiradiputra, M.R.D.; Thirapanmethee, K.; Khuntayaporn, P.; Wanapaisan, P.; Chomnawang, M.T. Comparative genotypic characterization related to antibiotic resistance phenotypes of clinical carbapenem-resistant *Acinetobacter baumannii* MTC1106 (ST2) and MTC0619 (ST25). *BMC Genom.* **2023**, *24*, 689.
9. Peleg, A.Y.; Seifert, H.; Paterson, D.L. *Acinetobacter baumannii*: Emergence of a successful pathogen. *Clin. Microbiol. Rev.* **2008**, *21*, 538–582. [PubMed]
10. Zhang, S.; Di, L.; Qi, Y.; Qian, X.; Wang, S. Treatment of infections caused by carbapenem-resistant *Acinetobacter baumannii*. *Front. Cell. Infect. Microbiol.* **2024**, *14*, 1395260.
11. Roberts, J.A.; Abdul-Aziz, M.-H.; Davis, J.S.; Dulhunty, J.M.; Cotta, M.O.; Myburgh, J.; Bellomo, R.; Lipman, J. Continuous versus intermittent β-lactam infusion in severe sepsis. A meta-analysis of individual patient data from randomized trials. *Am. J. Respir. Crit. Care Med.* **2016**, *194*, 681–691. [PubMed]

antibiotics

MDPI

Article

Acinetobacter baumannii Global Clone-Specific Resistomes Explored in Clinical Isolates Recovered from Egypt

Samira M. Hamed [1], Walid F. Elkhatib [2,3,*], Hanka Brangsch [4], Ahmed S. Gesraha [5], Shawky Moustafa [6], Dalia F. Khater [7], Mathias W. Pletz [8], Lisa D. Sprague [4], Heinrich Neubauer [4] and Gamal Wareth [4,8,*]

1 Department of Microbiology and Immunology, Faculty of Pharmacy, October University for Modern Sciences and Arts (MSA), Giza 12451, Egypt; satwa@msa.edu.eg

2 Microbiology and Immunology Department, Faculty of Pharmacy, Ain Shams University, African Union Organization Street, Cairo 11566, Egypt

3 Department of Microbiology and Immunology, Faculty of Pharmacy, Galala University, Suez 43727, Egypt

4 Institute of Bacterial Infections and Zoonoses, Friedrich-Loeffler Institut, 07743 Jena, Germany; hanka.brangsch@fli.de (H.B.); lisa.sprague@fli.de (L.D.S.); heinrich.neubauer@fli.de (H.N.)

5 Department of Pharmacology and Toxicology, Faculty of Pharmacy, Tanta University, Tanta 31511, Egypt; pg_171873@pharm.tanta.edu.eg

6 Faculty of Veterinary Medicine, Benha University, Toukh 13736, Egypt; shawky.mustafa@fvtm.bu.edu.eg

7 Tanta Laboratory, Animal Health Research Institute, Agricultural Research Center, Tanta 31511, Egypt; dr.daliakhater@gmail.com

8 Institute of Infectious Diseases and Infection Control, Jena University Hospital, 07747 Jena, Germany; mathias.pletz@med.uni-jena.de

* Correspondence: walid-elkhatib@pharma.asu.edu.eg (W.F.E.); gamal.wareth@fli.de (G.W.)

Abstract: *Acinetobacter baumannii* (*A. baumannii*) is a highly problematic pathogen with an enormous capacity to acquire or upregulate antibiotic drug resistance determinants. The genomic epidemiology and resistome structure of 46 *A. baumannii* clinical isolates were studied using whole-genome sequencing. The isolates were chosen based on reduced susceptibility to at least three classes of antimicrobial compounds and were initially identified using MALDI-TOF/MS, followed by polymerase chain reaction amplification of $bla_{\text{OXA-51-like}}$ genes. The susceptibility profiles were determined using a broth microdilution assay. Multi-, extensive-, and pan-drug resistance was shown by 34.8%, 63.0%, and 2.2% of the isolates, respectively. These were most susceptible to colistin (95.7%), amikacin, and trimethoprim/sulfamethoxazole (32.6% each), while only 26.1% of isolates were susceptible to tigecycline. In silico multi-locus sequence typing revealed 8 Pasteur and 22 Oxford sequence types (STs) including four novel STs (ST^{Oxf} 2805, 2806, 2807, and 2808). The majority of the isolates belonged to Global Clone (GC) 2 (76.4%), GC5 (19.6%), GC4 (6.5%), GC9 (4.3%), and GC7 (2.2%) lineages. An extensive resistome potentially conferring resistance to the majority of the tested antimicrobials was identified in silico. Of all known carbapenem resistance genes, $bla_{\text{OXA-23}}$ was carried by most of the isolates (69.6%), followed by IS*Aba1*-amplified bla_{ADC} (56.5%), $bla_{\text{NDM-1}}$ and $bla_{\text{GES-11}}$ (21.7% each), and $bla_{\text{GES-35}}$ (2.2%) genes. A significant correlation was found between carbapenem resistance and *carO* mutations, which were evident in 35 (76.0%) isolates. A lower proportion of carbapenem resistance was noted for strains possessing both $bla_{\text{OXA-23}}$- and $bla_{\text{GES-11}}$. Amikacin resistance was most probably mediated by *armA*, *aac(6′)-Ib9*, and *aph(3′)-VI*, most commonly coexisting in GC2 isolates. No mutations were found in *pmrABC* or *lpxACD* operons in the colistin-resistant isolates. Tigecycline resistance was associated with *adeS* (N268Y) and *baeS* (A436T) mutations. While the lineage-specific distribution of some genes (e.g., bla_{ADC} and $bla_{\text{OXA-51-like}}$ alleles) was evident, some resistance genes, such as $bla_{\text{OXA-23}}$ and *sul1*, were found in all GCs. The data generated here highlight the contribution of five GCs in *A. baumannii* infections in Egypt and enable the comprehensive analysis of GC-specific resistomes, thus revealing the dissemination of the carbapenem resistance gene $bla_{\text{OXA-23}}$ in isolates encompassing all GCs.

Keywords: *Acinetobacter baumannii*; WGS; global clones; resistance; resistome; Egypt; MDR; XDR; PDR

1. Introduction

Acinetobacter baumannii (*A. baumannii*) is among the most notorious and worrisome nosocomial pathogens currently receiving much attention [1]. Together with its extraordinary propensity to accumulate resistance genes, its distinct capacity for surviving harsh hospital environments and the very high potential of biofilm formation greatly contribute to the burden of hospital-acquired *A. baumannii* infections [2–4]. While mostly affecting intensive-care patients, reports of *A. baumannii* infections in other healthcare facilities are frequent [3]. Community-acquired cases of *A. baumannii* infections in patients with pre-existing comorbidities have also been reported [5,6].

Resistance to last-line antimicrobial agents has been globally encountered in *A. baumannii* infections [7,8]. Numerous reports on pan-drug-resistant *A. baumannii*, linked to high mortality, have been published [7,9,10]. Moreover, carbapenem-resistant *A. baumannii* was ranked by the World Health Organization (WHO) as the foremost pathogen for which new antimicrobial agents are urgently required [11]. Egypt is one of the countries most affected by carbapenem- and multidrug-resistant *A. baumannii* infections [12–16]. The isolation of carbapenem-resistant *A. baumannii* from foreigners returning from Egypt has been reported previously [17]. Deciphering the resistance background of *A. baumannii* isolates from Egypt could, therefore, add to our knowledge about this pathogen.

One aspect of antimicrobial resistance in *A. baumannii* that has not been thoroughly studied is the extent of variation between global clones. A clonal population structure has been identified for *A. baumannii* over the years [18], with at least nine successful globally disseminated clonal lineages known so far. These included global clones (GCs) 1 to 9 that are mostly dominated by GC2 and, to a lesser extent, GC1 [18–20]. An inter-clone variation of the pathogenic potential of *A. baumannii* has been demonstrated previously [18,21,22]. Likewise, a variation might also exist in their resistomes. Initially, the global clones only carried genes conferring resistance to aminoglycosides, sulfonamides, and tetracyclines. Due to various genetic mechanisms, sub-lineages also resistant to other antimicrobials, including carbapenems, could emerge successfully [21]. GC1 and GC2 have long been known to comprise the majority of isolates with multi-, extensive-, and pan-drug resistance phenotypes. Due to the dynamic nature and plasticity of the *A. baumannii* genome, a growing number of studies have reported the emergence of antimicrobial resistance phenotypes in other GCs as well [1,20,23,24]. Together with their pathogenic and epidemic potential, the resistome differences between apparently identical strains might reshape our perception of the risk posed by individual GCs and the public health precautions required for each. This might be facilitated by coupling the genomic epidemiology of *A. baumannii* isolates recovered in different parts of the world to their resistome structures.

Here, we conducted a large-scale whole-genome sequencing (WGS)-based study to explore the resistome structure of *A. baumannii* recovered from patients admitted to hospitals in Cairo, Egypt. In addition, we explored the contribution of different GCs to *A. baumannii* infections in Egypt and investigated their specific resistomes.

2. Results

2.1. Bacterial Isolates and Their Antimicrobial Susceptibility Profiles

Based on the quality of the WGS results, i.e., purity of sample material, only 46 of the 60 original *A. baumannii* isolates were further analysed. The metadata of the isolates are given in Supplementary Materials Table S2. Of all tested antimicrobial agents, the isolates were most often susceptible to colistin (95.7%). A lower proportion of isolates showed susceptibility to amikacin and trimethoprim/sulfamethoxazole (32.6%), while only 12 (26.1%) isolates were susceptible to TG. Even though all isolates were resistant to ertapenem, 15.2% and 13.0% of them were susceptible to imipenem and meropenem, respectively. Full or intermediate resistance was exhibited by all isolates to penicillins and the cephalosporins cefotaxime and ceftolozane/tazobactam. Other cephalosporins (ceftazidime, ceftazidime/avibactam, and cefepime) were inactive against the majority

(95.6–97.9%) of the isolates. Considering the full susceptibility profile of each isolate, 1 (2.2%), 16 (34.8%), and 29 (63.0%) isolates were classified as PDR, MDR, and XDR, respectively (Figure 1).

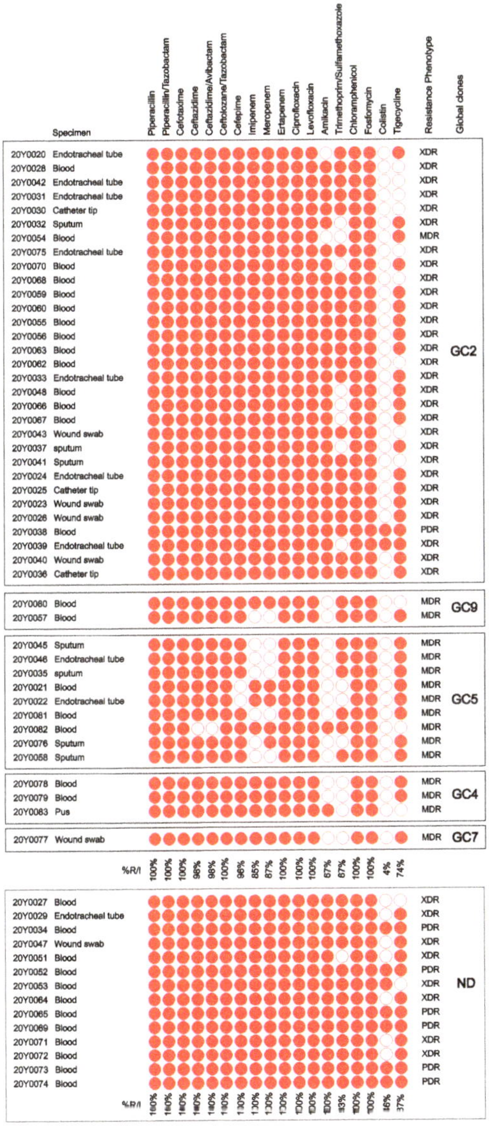

Figure 1. Clinical specimens and antimicrobial susceptibility profiles of all isolates. Reduced susceptibility (intermediate or resistance) is denoted by red icons, while the white ones correspond to full susceptibility. The isolates were segregated according to their global clones. MDR, multidrug-resistant; XDR, extensively drug-resistant; PDR, pan-drug resistant; GC, global clone; ND, the global clones could not be determined due to poor quality of draft genomes that were excluded from all subsequent genomic analyses; %R/I, percentage of reduced susceptibility to each antimicrobial agent. The figure was created using the MORPHEUS online tool (https://software.broadinstitute.org/morpheus, accessed on 3 January 2023) and edited using Inkscape version 1.2 (https://inkscape.org, accessed on 10 September 2022).

2.2. Draft Genomes and Annotation Features

Based on the short-read sequences, draft genomes consisting of 76 to 187 contigs were assembled with a mean N50 of 80,333 bp and covering 4approx. 81 to 86% of the reference genome (Table S3). The total length of the de novo assemblies varied between 3,798,591 bp and 4,247,432 bp with GC contents between 38.9% and 39.15%.

2.3. Phylogeny and Sequence Types

More than half of the isolates belonged to GC2 [25], encompassing four Pasteur STs and ten Oxford STs, as shown in Table S2. With less prevalence, one ST (ST602^{Pas}/ST732^{Oxf}) that belonged to GC5 [20] was also found in our collection, comprising 19.6% of the isolates. Three isolates (6.5%) belonged to GC4 [18] (ST15^{Pas}/ST1115^{Oxf}/). GC9 [19] was represented by two isolates that belonged to ST85^{Pas} and two Oxford STs (ST1580^{Oxf} and ST2808^{Oxf}). Only one isolate was found to belong to GC7 [26] (ST113^{Pas}/ST2246^{Oxf}). Additionally, MLST analysis of the isolates revealed four novel Oxford allelic profiles in eight isolates, to which novel Oxford STs were assigned. The novel STs comprised ST2805^{Oxf} (GC2), ST2806^{Oxf} (GC2), ST2807^{Oxf} (GC2), and ST2808^{Oxf} (GC9). Minimum spanning trees depicting Pasteur and Oxford STs are identified here and their clonal complexes are shown in Figures S1 and S2, respectively.

The cgSNP-based analysis encompassing the Egyptian isolates and global strains identified 64,994 SNPs. Five clusters that fully matched the GCs of the isolates were differentiated (Figure 2). GC2 isolates comprised the largest cluster (cluster 1), which was divided into two subclusters (1a and 1b). The larger one (cluster 1a) comprised all GC2 isolates except those belonging to ST281^{Oxf}. Members of cluster 1a showed the highest genomic diversity of all tested isolates. Up to 4012 SNP differences were found among cluster 1a isolates, with some showing higher similarity to foreign strains. Isolates 20Y0037, 20Y0041, and 20Y0043 differed by only four SNPs from strain A1819 isolated in Egypt in 2020. As few as 13, 16, 20, 29, and 31 SNPs separated several isolates of cluster 1a from foreign strains collected in the USA, Libya, France, Jordan, and Israel. Six ST208/1806^{Oxf} strains differed by only 13 SNPs from the US-American strain CFSAN, which dated back to 2005. Cluster 1b included ST281^{Oxf} isolates. These showed high similarity with an SNP range of 0 to 4. In contrast, the distance to strain ABUH763, which was collected in the USA in 2015, was very high (minimum SNPs = 647). Within cluster 2 harbouring GC9 isolates, isolate 20Y0057 showed a closer SNP distance (n = 265) to strain AbBAS-1 from Spain than strain M18 (352 SNPs) from Egypt isolated in 2020. Moreover, isolate 20Y0057 differed from isolate 20Y0080 collected in the current study by 692 SNPs. Only five SNPs were found in strain ACN-20190830-N60 isolated in Libya in 2019 compared with GC5 isolates collected in the current study. These, in turn, showed SNP differences ranging from 0 to 7. GC4 and GC5 contained the fourth and fifth cluster, respectively. Even though isolate 20Y0077 showed a relatively close SNP distance (SNPs = 19–25) to other Egyptian strains, it exhibited 12,290 SNPs compared with a strain from Libya collected in 2019.

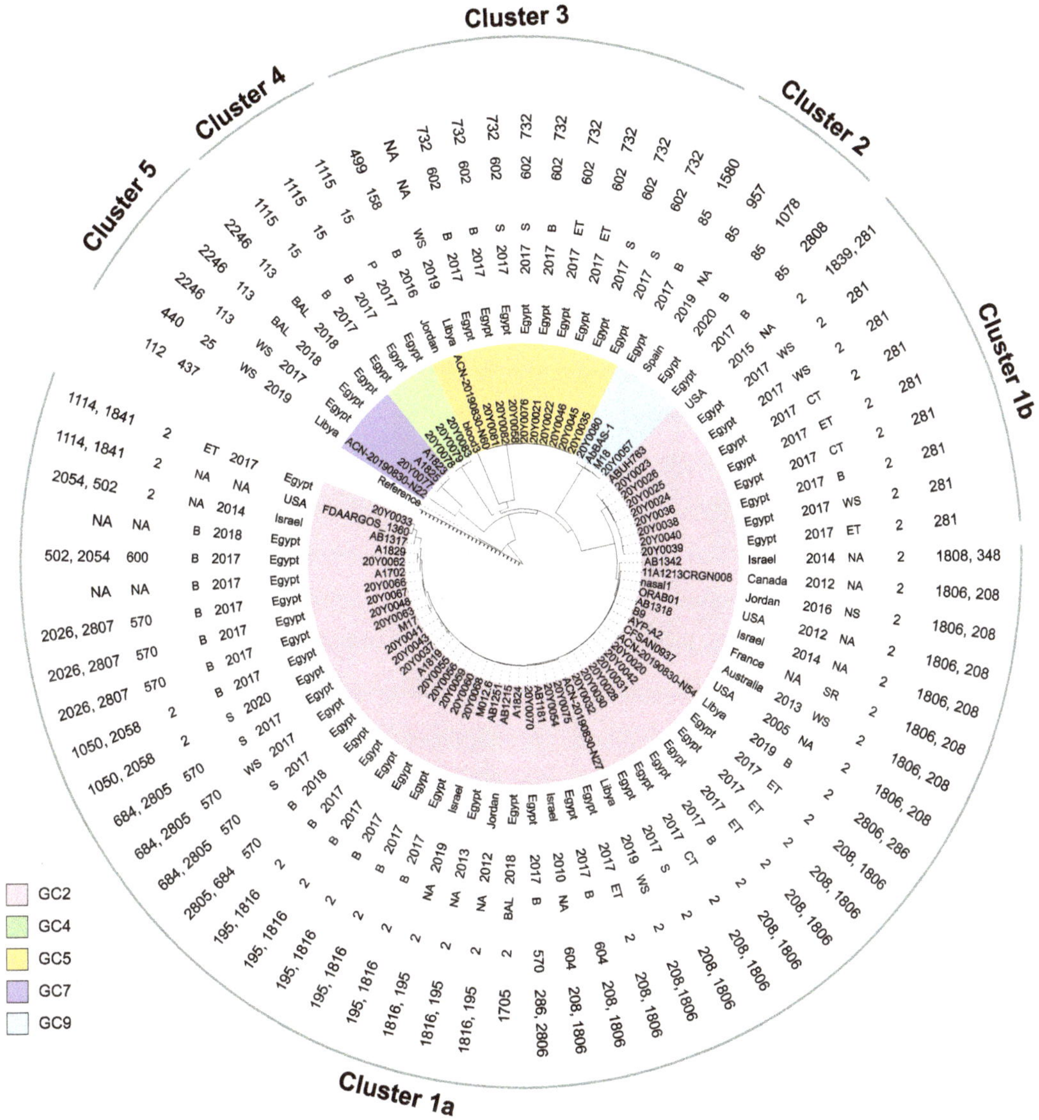

Figure 2. Maximum likelihood tree based on cgSNP alignment using *A. baumannii* ATCC 17978 (GCF_001593425.2) as the reference genome. Concentric labels refer to (from inside to outside): the country of isolation, year of isolation, specimen type, ST$^{(Pas)}$, and ST$^{(Oxf)}$. Clusters are coloured according to their global clones (GCs). B, blood; S, sputum; P, pus; WS, wound swab; ET, endotracheal tube; CT, catheter tip; BAL, bronchoalveolar lavage; and SR, Seine River. The tree was visualized using the interactive tree of life (iTOL) online tool v6.7 (https://itol.embl.de/, accessed on 12 March 2023) and edited using Inkscape version 1.2 (https://inkscape.org, accessed on 10 September 2022).

2.4. Correlation between Antimicrobial Resistance and Global Clones

Compared with the other investigated isolates, the isolates that belonged to GC2 showed higher resistance to amikacin (93.5% vs. 13.3%), colistin (6.5% vs. 0.0%), trimethoprim/sulfamethoxazole (74.2% vs. 53.3%), ceftazidime (100.0 vs. 93.3%), ceftazidime/avibactam (100.0 vs. 93.3%), imipenem (100.0% vs. 53.3%), and meropenem (100.0% vs. 60.0%).

This was statistically significant only for amikacin, imipenem, and meropenem with p-values < 0.001. However, resistance to TG was lower in GC2 isolates than in the isolates belonging to other GCs (71.0% vs. 80.0%, p-value = 0.723). The XDR phenotype was exclusively found in GC2 isolates (p-value < 0.001).

2.5. Resistome Structure and Its Correlation to the Resistance Phenotypes

Scanning the draft genomes for antimicrobial resistance genes revealed a large resistome comprising at least 51 resistance determinants known to potentially confer resistance to ten antimicrobial classes through five resistance mechanisms, shown in Figure 3. Of these, antibiotic enzyme inactivation prevailed, followed by drug efflux, target alteration, target replacement, and finally target protection. Inactivating enzymes most frequently targeted aminoglycosides and β-lactams, including carbapenems. Detailed descriptions of the resistome and the potential contribution of different determinants to antimicrobial resistance exhibited by our collection are also shown in Figure 3.

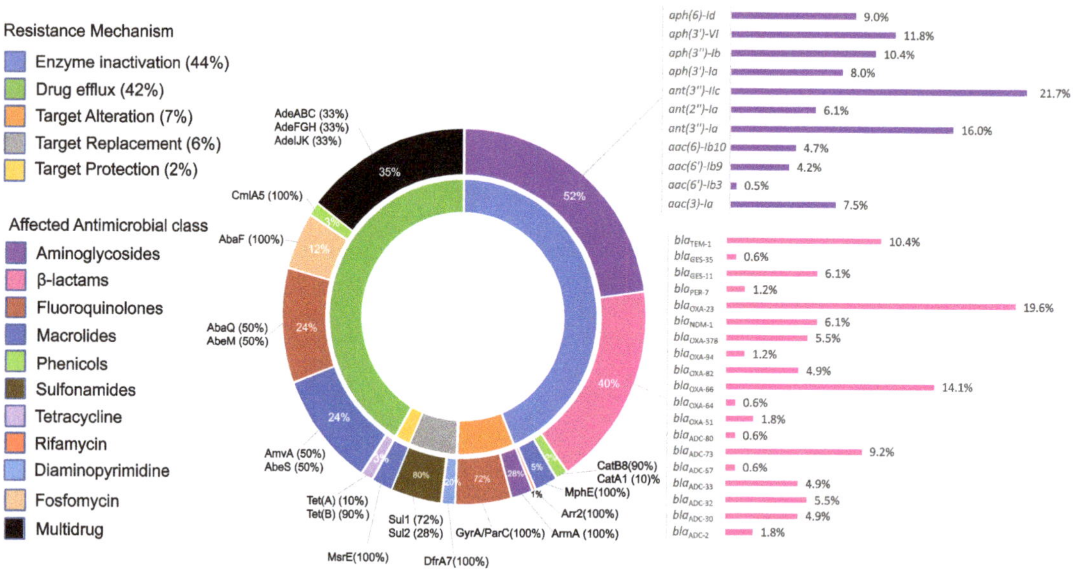

Figure 3. Graphical representation of the in silico identified potential resistance determinants of the total collection of *A. baumannii* isolates. The figure shows the contribution of different resistance mechanisms to the resistome and the antimicrobial classes affected by each. For each antimicrobial class, the contribution of individual resistance determinants is also shown. The figure was created using Microsoft Office Excel 2016 and edited using Inkscape version 1.2 (https://inkscape.org, accessed on 10 September 2022).

A minimum of eleven resistance determinants were carried by all isolates. These included the efflux pump-coding genes *ade*ABC, *ade*FGH, *ade*IJK, *aba*F, *aba*Q, *amv*A, *abeM*, and *abe*S; the intrinsic aminoglycoside modifying enzyme-coding gene *ant(3″)-IIc* [27]; and, finally, six variants of the $bla_{OXA51\text{-like}}$ genes and seven alleles of the *Acinetobacter*-derived cephalosporinase (ADC)-coding genes conferring intrinsic resistance to β-lactams [28,29].

As inferred by Resfinder, the acquired genes conferring resistance to the last-line antimicrobials commonly used for treating MDR *A. baumannii* infections are detailed below. In addition, resistance to other antimicrobial classes was also acquired by our isolates, as shown in Figure 4. Most frequently, they targeted fluoroquinolones (100%) through mutations affecting the quinolone resistance-determining regions (QRDRs) of *gyrA* and *parC*. At least one sulfonamide/diaminopyrimidine resistance gene was carried by 71.7% of the isolates. Genes with the potential to confer resistance to chloramphenicol, macrolides, tetracyclines, and rifamycin were detected in 47.8%, 41.3%, 21.7%, and 6.5% of the isolates, respectively. Up to 17 acquired resistance determinants coexisted in our isolates.

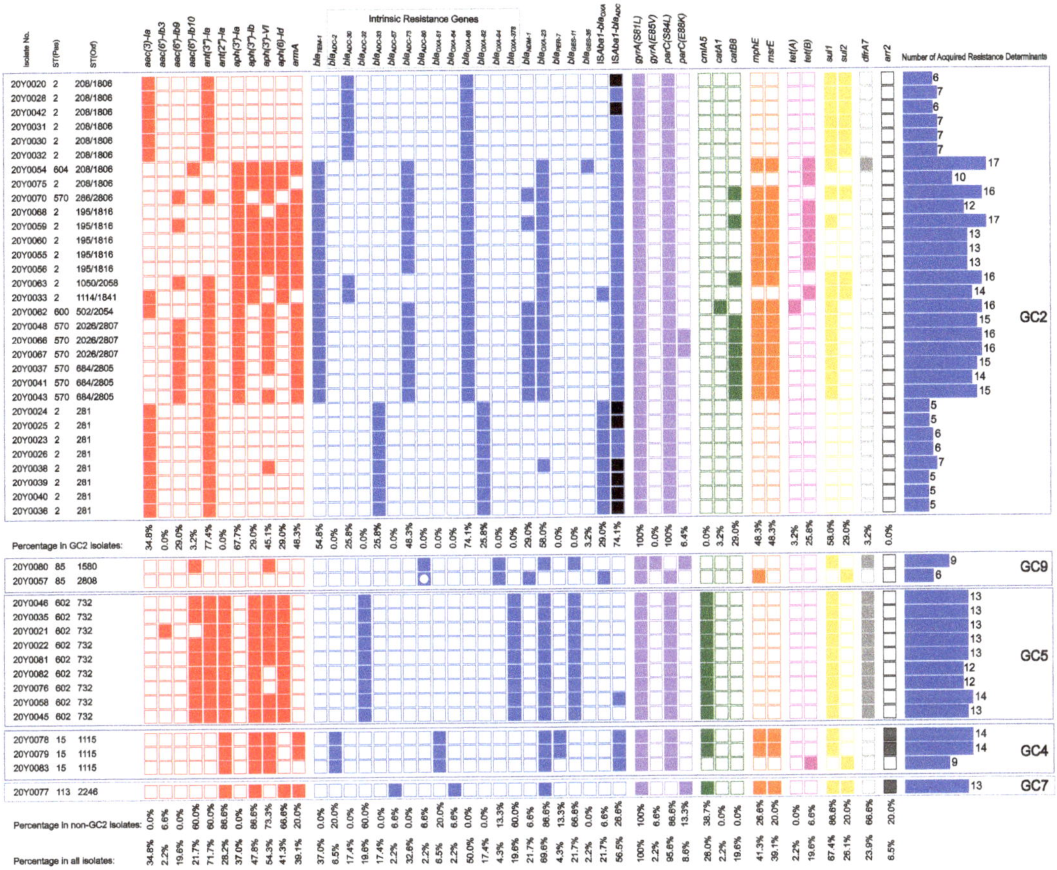

Figure 4. Distribution of resistance determinants potentially conferring resistance to different antimicrobial classes in all isolates. The isolates were arranged according to their phylogenetic relationships and were grouped into global clones. The presence or absence of resistance determinants is denoted by coloured or white squares, respectively. The white circle within the blue square denotes an interrupted *bla*$_{ADC}$ gene. The figure is divided into panels of icons with different colours corresponding to antimicrobial classes affected by different resistance determinants. Red, aminoglycosides; blue, β-lactams; purple, fluoroquinolones; green, phenicoles; deep rose, tetracyclines; yellow, sulfonamides; grey, diaminopyrimidine; and dark grey, rifamycin. Undetermined IS*Aba1* upstream to *bla*$_{OXA}$ or *bla*$_{ADC}$ is denoted by black squares. The total numbers of acquired resistance genes carried by each isolate are represented by blue bars. The figure was created using the iTOL online tool v6.7 (https://itol.embl.de/, accessed on 12 March 2023) and edited using Inkscape version 1.2 (https://inkscape.org, accessed on 10 September 2022).

2.5.1. Carbapenem Resistance

In addition to the genes coding narrow (*bla*$_{TEM-1}$) and extended-spectrum (*bla*$_{PER-7}$) β-lactamases affecting penicillins and different generations of cephalosporins, carbapenem resistance determinants were also abundantly present in our collection. Of all known carbapenem resistance genes, *bla*$_{OXA-23}$ prevailed (69.6%). Less frequently, the isolates carried IS*Aba1*-amplified *bla*$_{ADC}$ (56.5%), *bla*$_{NDM-1}$ and *bla*$_{GES-11}$ (21.7% each), and finally *bla*$_{GES-35}$ (2.2%). These were more likely to co-exist in combinations of two (50.0%) or three determinants (23.9%), as shown in Table 1. All except *bla*$_{GES-11}$ were more frequently carried by carbapenem-resistant isolates compared with others. This was significant only for IS*Aba1*-preceded *bla*$_{ADC}$ (Fisher's Exact Test: p-value = 0.015 for imipenem and 0.036 for meropenem). The association between combinations of carbapenemase-coding genes and carbapenem resistance is shown in Table 1.

Table 1. Correlation between potential carbapenem resistance determinants and carbapenem resistance.

Carbapenem Resistance Determinants	No. of Isolates	GC	IMP^R	MER^R
None	2	2	100.0%	100.0%
IS*Aba1*-preceded *bla*$_{OXA-51-like}$ only	5	2	100.0%	100.0%
IS*Aba1*-preceded *bla*$_{ADC}$ only	4	2	100.0%	100.0%
bla$_{OXA-23}$ only	1	7	100.0%	100.0%
IS*Aba1*-preceded *bla*$_{OXA-51-like}$ + IS*Aba1*-preceded *bla*$_{ADC}$	2	2	100.0%	100.0%
bla$_{OXA-23}$ + IS*Aba1*-preceded *bla*$_{OXA-51-like}$	1	2	100.0%	100.0%
bla$_{OXA-23}$ + IS*Aba1*-preceded *bla*$_{ADC}$	9	2 and 4	100.0%	100.0%
bla$_{NDM-1}$ + IS*Aba1*-preceded *bla*$_{OXA-51-like}$	1	9	0.0%	0.0%
bla$_{OXA-23}$ + *bla*$_{NDM-1}$	1	2	100.0%	100.0%
bla$_{OXA-23}$ + *bla*$_{GES-11}$	9	5 and 9	44.4%	55.6%
bla$_{OXA-23}$ + IS*Aba1*-preceded *bla*$_{OXA-51-like}$ + IS*Aba1*-preceded *bla*$_{ADC}$	1	2	100.0%	100.0%
bla$_{OXA-23}$ + *bla*$_{NDM-1}$ + IS*Aba1*-preceded *bla*$_{ADC}$	8	2	100.0%	100.0%
bla$_{OXA-23}$ + *bla*$_{GES-11}$ + IS*Aba1*-preceded *bla*$_{ADC}$	1	5	0.0%	0.0%
bla$_{OXA-23}$ + *bla*$_{GES-35}$ + IS*Aba1*-preceded *bla*$_{ADC}$	1	2	100.0%	100.0%

IMP^R, imipenem resistance and MER^R, meropenem resistance.

The contribution of the carbapenem susceptibility porin (CarO) to carbapenem resistance was also analysed. For this purpose, the nucleotide sequence of *carO* was extracted from all draft genomes. Compared with the wildtype gene sequence of *A. baumannii* ATCC 19606, multiple mutations were identified. The wildtype sequence of *carO* was fully retained by only 11 isolates comprising the GCs 5 and 9. Susceptibility to carbapenems was significantly more prevalent among the isolates carrying the wildtype genes (Fisher's Exact Test, p-value < 0.001). None of the isolates carrying the mutated *carO* gene (35 isolates) were susceptible to any of the tested carbapenems. Multiple sequence alignment of the predicted amino acid structure of the protein product of the *carO* gene from all isolates compared with that of *A. baumannii* ATCC 19606 is shown in Figure S3.

2.5.2. Amikacin Resistance

As shown in Figure 3, the aminoglycoside-modifying enzymes (AMEs) had the highest proportion among the genes encoding antimicrobial-inactivating enzymes found in the isolates. At least 11 acquired AME-coding genes were detected. Specifically, those affecting amikacin were thoroughly analysed. Only *aph(3′)-VI* and *aac(6′)-Ib* are known to affect amikacin. Corresponding genetic determinants were found in 54.3% and 43.5% of the isolates, respectively. An additional enzyme conferring broad-spectrum aminoglycoside resistance is the 16S rRNA methyltransferase enzyme coded by *armA*. The gene was identified in 39.1% of the isolates spanning three GCs (GC2, GC4, and GC7). Notably, the three genes coexisted exclusively in GC2 isolates. This corresponded to the significantly higher amikacin resistance recorded for GC2 isolates compared with others (93.5% versus

13.3%, p-value < 0.001). The combinations of amikacin resistance genes, their prevalence in different global clones, and their correlation to amikacin resistance are shown in Table 2.

Table 2. Amikacin resistance genes found in the Egyptian *A. baumannii* isolates and their correlation to observed amikacin resistance.

Aminoglycoside Resistance Determinants	No. of Isolates	GC	AMK^R
None	15	2 and 9	86.7%
aph(3′)-VI only	3	2 and 4	100.0%
aac(6′)-Ib10 only	2	5	50.0%
armA only	2	2 and 7	50.0%
armA + *aac(6′)-Ib9*	2	2	100.0%
armA + *aph(3′)-VI*	6	2 and 4	66.7%
aph(3′)-VI + *aac(6′)-Ib10*	7	5 and 9	0.0%
aph(3′)-VI + *aac(6′)-Ib3*	1	5	0.0%
aph(3′)-VI + *aac(6′)-Ib10* + *armA*	1	2	0.0%
aph(3′)-VI + *aac(6′)-Ib9* + *armA*	7	2	100.0%

AMK^R, amikacin resistance.

Amikacin resistance was more common among the isolates carrying *armA* (77.8%) and *aph(3′)-VI* (56.0%) compared with others. Among the isolates with the potential for *aac(6′)-Ib* production, only those carrying *aac(6′)-Ib9* were non-susceptible to amikacin (Fisher's Exact Test, p-value = 0.018). The association of *aph(3′)-VI* with an upstream IS*Aba125*, through which overexpression is known to be mediated [30], was also analysed. The analysis showed that *aph(3′)-VI* carried by all bla_{NDM-1}-positive GC2 isolates was preceded by IS*Aba14* rather than IS*Aba125*. Interestingly, none of these were susceptible to amikacin.

2.5.3. Colistin Resistance

Only two isolates (20Y0037 and 20Y0066) from our collection were non-susceptible to colistin. Both belonged to GC2, ST570^{Pas}, and a novel Oxford ST (684/2805^{Oxf} for 20Y0037 and 2026/2807^{Oxf} for 20Y0066). To explore the potential mechanisms underlying the colistin resistance in the two isolates, the *pmrCAB* and *lpxACD* operons were analysed. Using *A. baumannii* ATCC 19606 as reference, the gene sequences exhibited multiple missense mutations that resulted in amino acid alterations in *pmr*C (V58I—F166L—N300D—A370S—K531T), *pmr*B (A138T—N440H—A444V), *lpx*C (N287D), and *lpx*D (E117K). Similar analysis of other colistin-susceptible isolates with the same STs revealed the same mutation pattern, which was therefore considered a polymorphism. No unique genes were identified in the colistin-resistant isolates compared to other isolates with the same sequence type.

2.5.4. Tigecycline Resistance

A total of 34 TG-resistant isolates was identified in our collection. Mutations previously linked to TG resistance [31–34] were carefully analysed. None of the TG-resistant isolates carried target site mutations in the ribosomal protein S10-coding gene (*rpsJ*). Mutations affecting the regulatory genes of the *ade*ABC efflux system, for which TG is a well-known substrate, were also analysed. These include *ade*RS and *bae*SR. Multiple mutations were found in the two regulatory systems carried by all strains when compared with *A. baumannii* ATCC 19606. Multiple sequence alignments of the predicted amino acid sequences of *ade*R, *ade*S, *bae*R, and *bae*S are shown in Figures S4–S7. Mutation hotspots of all genes and their correlation to TG resistance are also summarised in Table S4. Mutation patterns were found to be correlated with GCs rather than TG susceptibility profiles. With few exceptions, no mutations were found in TG-resistant isolates compared to TG-susceptible isolates of the same ST/GC. Nonetheless, one missense mutation in *adeS* resulting in the amino acid alteration N268Y was associated with TG resistance in all affected isolates. Similarly, threonine replaced alanine in position 436 of BaeS in the TG-resistant isolate 20Y0057 compared with the TG-susceptible isolate 20Y0080 which belonged to the same GC. While missense mutations were found in *adeRS* and *baeRS* carried by the TG-resistant isolate 20Y0077 (the solitary GC7 isolate) compared with *A. baumannii* ATCC19606, these were also found in closely related genomes retrieved from the BV-BRC database. While recently linked to TG resistance, *tet(A)* was carried by only one isolate in our collection that retained susceptibility to TG.

2.6. Global Clone-Specific Resistomes

As shown in Figure 4, the average number of acquired resistance determinants was highest (14.6) among a subset of GC2 isolates possessing both *bla*$_{OXA-23}$ and *bla*$_{TEM-1}$. Moreover, fewer determinants were carried by isolates of GC7 (13), GC5 (12.8), GC4 (12.3), and GC9 (7.5) than those belonging to GC2 that did not possess *bla*$_{OXA-23}$ and *bla*$_{TEM-1}$ (6).

In addition to the GC-specific mutations identified in the efflux pump regulatory genes described above, uneven distribution of some resistance determinants was also evident in different GCs (Figure 5). Some genes were unique for specific GCs, while others were common to some or all GCs. Additionally, some combinations of carbapenemase resistance determinants and amikacin resistance determinants were also unique for specific GCs, as shown in Tables 1 and 2. Only two resistance genes were found to be present in all GCs (Figure 5). These included the carbapenemase-coding gene *bla*$_{OXA-23}$ and the sulfonamide resistance gene *sul1*.

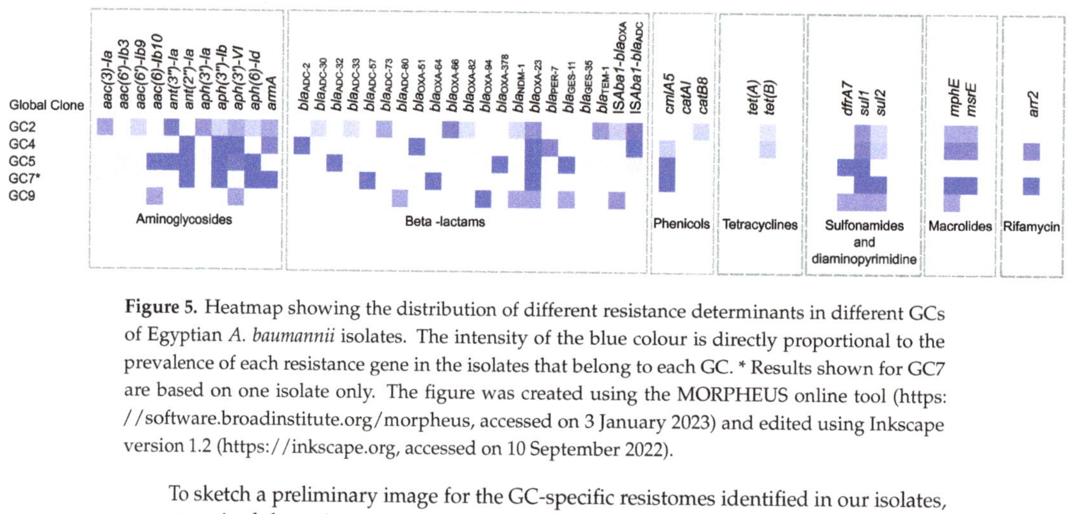

Figure 5. Heatmap showing the distribution of different resistance determinants in different GCs of Egyptian *A. baumannii* isolates. The intensity of the blue colour is directly proportional to the prevalence of each resistance gene in the isolates that belong to each GC. * Results shown for GC7 are based on one isolate only. The figure was created using the MORPHEUS online tool (https://software.broadinstitute.org/morpheus, accessed on 3 January 2023) and edited using Inkscape version 1.2 (https://inkscape.org, accessed on 10 September 2022).

To sketch a preliminary image for the GC-specific resistomes identified in our isolates, we categorised the resistance genes into three groups: unique genes (found exclusively in one GC), conserved genes (found in all isolates of one or more GCs), and finally genes that were present but not fully conserved in one or more GCs (Table 3). Note that GC7 was not included in this analysis, as it was represented by a single isolate.

Table 3. Categories of antimicrobial resistance genes detected in Egyptian *A. baumannii* isolates based on their occurrence in different GCs.

GC	Unique	Fully Conserved	Not Fully Conserved
bla_{OXA-23} / bla_{TEM-1}-positive GC2	*aph(3′)-Ia*/bla_{TEM-1}/bla_{ADC-73}/bla_{GES-35}/*catA1*/*catB8*/*tetA*	*aph(3′)-Ia*/bla_{TEM-1}/bla_{OXA-66}/bla_{OXA-23}/IS*Aba1*-bla_{ADC}	*aac(3)-Ia*/*aac(6′)-Ib*/*ant(3″)-Ia*/*aph(3″)-Ib*/*aph(3′)-VI*/*aph(6)-Id*/*armA*/bla_{ADC-30} */bla_{ADC-73}/bla_{NDM-1}/bla_{GES-35}/IS*Aba1*-$bla_{OXA-51-like}$/*catA1*/*catB8*/*mphE*/*msrE*/*tetA*/*tetB*/*sul1*/*sul2*/*dfrA7*
bla_{OXA-23} / bla_{TEM-1}-negative GC2	bla_{ADC-33}/bla_{OXA-82}		*aac(3)-Ia*/*ant(3″)-Ia*/*aph(3′)-VI*/bla_{ADC-30} */bla_{ADC-33}/bla_{OXA-66}/bla_{OXA-82}/bla_{OXA-23}/IS*Aba1*-$bla_{OXA-51-like}$/IS*Aba1*-bla_{ADC}/*sul1*/*sul2*
GC4	bla_{ADC-2}/bla_{OXA-51}/bla_{PER-7}	*ant(2″)-Ia*/*aph(3″)-Ib*/*aph(3′)-VI*/bla_{ADC-2}/bla_{OXA-51}/bla_{OXA-23}/IS*Aba1*-bla_{ADC}	*armA*/bla_{PER-7}/*cmlA5*/*mphE*/*msrE*/*tetB*/*sul1*/*sul2*/*arr2*
GC5	bla_{ADC-32}/$bla_{OXA-378}$	*aac(6′)-Ib*/*ant(3″)-Ia*/*ant(2″)-Ia*/*aph(3″)-Ib*/*aph(6)-Id*/bla_{ADC-32}/$bla_{OXA-378}$/bla_{OXA-23}/bla_{GES-11}/*cmlA5*/*sul1*/*dfrA7*	*aph(3′)-VI*/IS*Aba1*-bla_{ADC}
GC9	bla_{ADC-32}/bla_{OXA-94}	bla_{OXA-94}	*aac(6′)-Ib*/*aph(3′)-VI*/bla_{ADC-32}/bla_{NDM-1}/bla_{OXA-23}/bla_{GES-11}/IS*Aba1*-$bla_{OXA-51-like}$/*mphE*/*sul1*/*sul2*/*dfrA7*

*, the gene is unique for GC2.

3. Discussion

The present study explored the genomic epidemiology of MDR, XDR, and PDR *A. baumannii* isolates recovered from tertiary care hospitals in Egypt. The resistome of the isolates and the global clone-specific resistomes were explored in vitro and using WGS. The phenotypic analysis of the isolates showed that the majority were XDR (63.0%), while 34.8% were MDR. Only one isolate showed a PDR phenotype. MDR and XDR *A. baumannii* have been abundantly isolated from Egyptian patients in previous studies [12,15,16,35,36]. A study from Egypt reported the isolation of PDR *A. baumannii* from a patient with ventilator-associated pneumonia; however, the authors did not test the susceptibility of the isolate to tigecycline [37]. In our collection, the highest susceptibility was shown for colistin, followed by amikacin and trimethoprim/sulfamethoxazole. Colistin resistance in carbapenem-resistant *A. baumannii* was first reported in Egypt by Abdulzahra et al. in 2018 [38]. While a high prevalence of colistin resistance was reported by Fam et al. [39], comparatively low levels of resistance were found among the present collection of isolates and by other studies from Egypt [12,13,15]. Similar susceptibility levels were shown by our isolates to both amikacin and trimethoprim/sulfamethoxazole. Moreover, trimethoprim/sulfamethoxazole showed superior activity against GC2 isolates compared with amikacin. The overall resistance to amikacin found here was comparable to the regional levels [13,15]. The relatively high susceptibility to trimethoprim/sulfamethoxazole supports the recent calls for reviving old antimicrobials such as trimethoprim/sulfamethoxazole, alone or in combination with colistin, for treatment of carbapenem-resistant *A. baumannii* infections [40,41]. Contrary to our findings, a higher level of resistance to trimethoprim/sulfamethoxazole was reported by other studies from Egypt [12,15].

The resistance of our isolates to penicillins and cephalosporins ranged from 96 to 100%. Due to their intrinsic class C and D β-lactamases, the majority of *A. baumannii* strains isolated locally [13,15] or globally [42,43] are non-susceptible to penicillins and cephalosporins. Of these, the highest activity against our isolates was shown by cefepime, the least affected by class C β-lactamases [44]. Up to 87% of the isolates were non-susceptible to imipenem/meropenem. This is comparable to the results of other studies from Egypt [13,45,46]. Meanwhile, all of the isolates were non-susceptible to ertapenem, which is known for its limited activity against glucose-non-fermenting Gram-negative bacteria including *A. baumannii* [47–49]. No activity was found for the tested fluoroquinolones against any of our isolates. Similar findings have also been reported by others [12,45]. In contrast to previous findings [12,15,50], a high number of TG-resistant isolates were found in our collection. Fosfomycin and chloramphenicol are not listed among the antimicrobial agents recommended by the CLSI for routine susceptibility testing in *A. baumannii* infections. Nevertheless, some studies have demonstrated their synergistic activity with colistin against *A. baumannii* [51–53]. The susceptibility of our isolates to both agents was investigated to assess potential effectiveness in treating MDR *A. baumannii* infections. However, all isolates were non-susceptible to both agents.

All STs assigned to the isolates included in our study were found to belong to high-risk GCs (GC2, GC4, GC5, GC7, and GC9), most of which have been previously reported in Egypt by other studies [12,15,39,54]. Although GC1 predominated in one study [12], none of our isolates belonged to this GC. However, GC2 isolates prevailed in our collection, as observed in previous studies from Egypt [15,39,54,55]. The prevalence of GC2 isolates among multidrug-resistant *A. baumannii* has been also reported globally [20,56–58]. A considerable share of our isolates had the $ST602^{Pas}/732^{Oxf}$ and, as determined by Al-Hassan et al. [20], belonged to GC5. Notably, this ST was first identified in *A. baumannii* isolates originating from Egypt [46] and was later reported from other countries such as Libya [59], Tunisia [60], Jordan [61], Germany [62], and Sudan [20]. Moreover, the cgSNP-based analysis of the isolates revealed a close evolutionary relationship to international strains, reflecting the great potential of the STs identified here for spreading across continents.

An extensive resistome comprising almost all known resistance mechanisms affecting several classes of antimicrobials was identified in silico in our collection. Unexpectedly, no plasmid-mediated quinolone resistance genes were identified [63,64]. However, all isolates carried mutations in the QRDRs of *gyrA* and *parC* genes. A highly conserved mutation pattern was observed in the isolates with only a few exceptions. While the *gyrA* mutation associated with S81L amino acid alteration was fully conserved in all isolates, some variability was found in *parC* mutations. The majority (91.3%) of isolates displayed the mutation pattern $gyrA^{S81L}/parC^{S84L}$. This mutation pattern has been linked to high-level fluoroquinolone resistance in *A. baumannii* [63,65,66]. However, some isolates showed different mutation patterns. In addition to the $gyrA^{S81L}$ mutation, the isolates 20Y0066 and 20Y0067 carried two *parC* mutations, resulting in the amino acid alterations S84L and E88K. In agreement with [67], a single mutation in codon 88 of *parC* (E88K) in association with the $gyrA^{S81L}$ mutation could be the reason for fluoroquinolone resistance in isolate 20Y0077. Isolate 20Y0080 carried a single *parC* mutation (E88K) and two mutations in *gyrA* (S81L and E85V). Mutations affecting codon 85 of *gyrA* have been observed in ciprofloxacin-resistant *A. baumannii* [67]. In addition to target site alterations, fluoroquinolones are extruded by a variety of efflux pumps whose coding genes were carried by all of our isolates. These included *adeABC*, *adeIJK*, *adeFGH*, *abeM*, and *abeS*.

Resistance to chloramphenicol shown by all of our isolates was conferred by the efflux pumps *ade*ABC, *ade*IJK, *ade*FGH, and *abe*M [68], whose coding genes were carried by all isolates. Only a small fraction of the isolates carried the chloramphenicol acetyltransferase-coding genes *catA4* and *catB5* and the major facilitator superfamily (MFS) efflux pump-coding gene *cmlA5*. Likewise, fosfomycin resistance was probably conferred by the major facilitator superfamily (MFS) transporter AbaF. Target site alteration was the major resistance mechanism for the antifolate combination of trimethoprim and sulfamethoxazole. Here, we identified *dfrA7*, which encoded a trimethoprim-resistant dihydrofolate reductase in 23.9% of the isolates, predominantly in GC5. Additionally, the sulfonamide-resistant dihydropteroate synthase-encoding genes *sul1* and *sul2* were carried by 67.4% and 26.1% of the isolates, respectively. Notably, they were distributed in all GCs.

Carbapenems have been the most appropriate treatment option for *A. baumannii* infections until the significant rise of carbapenem resistance throughout the world [69–71]. The bla_{OXA-23}, the first known oxacillinase with carbapenem hydrolysing capacity [72] and the major contributor to carbapenem resistance in *A. baumannii* [73], was the most common (69.7%) carbapenem resistance determinant identified in our collection across all GCs. In line with this finding, a high prevalence has been reported from Egypt by other authors [74–76]. Most commonly, bla_{OXA-23} was combined with other potential carbapenem resistance genes. IS*Aba1*-amplified intrinsic bla_{OXA}- and bla_{ADC}-type β-lactamase-coding genes affecting carbapenems [77] were also found with a high prevalence. While IS*Aba1*-preceded-$bla_{OXA-51-like}$ genes were confined to GC2 and GC9, IS*Aba1*-bla_{ADC} was found in GC2, GC4, and GC5. IS*Aba1*-amplified $bla_{OXA-51-like}$ genes likely compensated for the lack of the bla_{OXA-23} gene in ST281^{Oxf}/bla_{OXA-82} carbapenem-resistant isolates in which they were identified as the sole known carbapenem resistance mechanism. This has also been reported by Zander et al. [78].

Metallo-β-lactamases have been known to possess a 100–1000 fold higher carbapenemase activity than OXA-type carbapenemases [18]. Of these, bla_{NDM-1} predominated in most studies conducted on *A. baumannii* [15,56,74,76], and in our collection, too. Here, the gene bla_{NDM-1} was exclusively carried by some GC2 and GC9 isolates. The bla_{NDM-1} production was previously linked to ST85 (GC9) [18], which was thought to be a reservoir for bla_{NDM-1} in the Middle East [16,20,79]. With the same prevalence as bla_{NDM-1}, the potential for Guyana extended spectrum β-lactamase variant 11 (GES-11) production was also evident in our collection. However, the weak potential of GES-11 for the hydrolysis of carbapenems was previously documented by [80]. The authors reported an imipenem MIC of 1 μg/mL for the bla_{GES-11} *A. baumannii* transconjugant compared with MIC of >24 μg/mL for the transconjugant carrying bla_{OXA-23}. A lower imipenem resistance was noted in our collection among the group of isolates in which bla_{OXA-23} coexisted with bla_{GES-11} compared with those carrying bla_{OXA-23} with or without other carbapenem resistance genes. Despite the low sequence homology between the OXA-23 and GES-11, the two enzymes were previously reported to be anchored in the periplasmic space of *A. baumannii* AB5075 by interacting with the same lysine sites of the periplasmic domain of the outer membrane protein A (OmpA) [81]. The impact of this competitive binding to OmpA on the carbapenemase activity of OXA-23 might be of importance. Further studies are required to support this finding. Notably, bla_{GES-11} was confined to GC5 isolates and one GC9 isolate, and the allele bla_{GES-35} coexisted with bla_{OXA-23} in one carbapenem-resistant isolate. GES production has been observed in *A. baumannii* isolates from Egypt [13,76,82]. While previously linked to carbapenem resistance in *A. baumannii* isolates from Egypt [74,75,82], genes for *Klebsiella pneumoniae* carbapenemase (KPC) were not found in any of our isolates.

The influx of imipenem in *A. baumannii* has been known to be partly mediated by the outer membrane channel protein CarO [83]. Hence, amino acid alterations in CarO have been implicated in imipenem resistance. While overlooked by most studies, mutations affecting the CarO protein and their association with carbapenem resistance were investigated in the current study. A significant association was found between carbapenem resistance and *carO* mutations, corroborating previous studies [84,85].

Amikacin is one of the few antimicrobials for which susceptibility was retained by a substantial fraction of *A. baumannii* strains, isolated locally [15] and globally [86]. Among other aminoglycosides, resistance to amikacin is known to be imparted by *armA*, carried by 39.1% of the isolates. Additionally, numerous genes coding for aminoglycoside-modifying enzymes with the potential to inactivate different aminoglycosides, including amikacin (Table 2), were identified in our collection. Most of the isolates carrying either *armA*, *aph(3′)-VI*, or *aac(6′)-Ib9* were non-susceptible to amikacin. This was statistically significant only for *aac(6′)-Ib9*. However, 13/15 isolates without the mentioned genes were resistant to amikacin, highlighting the significant contribution of the efflux pumps in aminoglycosides resistance and the contribution of other resistance mechanisms that were hitherto unknown. High-level amikacin resistance due to *aph(3′)-VI* has been previously linked to IS*Aba125*-mediated overexpression of *aph(3′)-VI* (*aphA6*) [30,87]. As reported by others [88,89], we found that the gene was preceded by IS*Aba14* rather than IS*Aba125* in some isolates that were all non-susceptible to amikacin. This suggests that IS*Aba14* may have the same impact on *aph(3′)-VI* expression as IS*Aba125*. The synergistic effect of the amikacin resistance mechanism was also indicated by the significantly higher prevalence of amikacin resistance in GC2 isolates that carried combinations of *armA*, *aph(3′)-VI*, and *aac(6′)-Ib*.

Colistin and TG are currently the last resort for the treatment of multidrug resistant *A. baumannii* infections [90]. Colistin resistance has been linked to either the complete loss or modification of the outer membrane lipopolysaccharides, as well as to the presence of the mobilised colistin resistance (*mcr*) gene. The former impairs the membrane's integrity, thereby potentiating susceptibility to other antimicrobial agents. This is mediated by mutations affecting lipid A biosynthesis genes (*lpxACD*) [91]. More commonly, colistin resistance arises due to the modification of bacterial lipopolysaccharides resulting from *pmrCAB* mutations [92]. Only two isolates in our collection of novel Oxford sequence types were non-susceptible to colistin. While missense mutations were found in the genes *pmrB*, *pmrC*, *lpxC*, and *lpxD* carried by the two strains compared with the corresponding genes of *A. baumannii* ATCC 19606, they were also found in colistin-susceptible isolates that belonged to the same STs. The altered expression of *pmrC* under the control of unexplored regulatory systems or another unknown colistin resistance mechanism might be the reason.

Although TG was specifically derived from tetracycline to evade tetracycline resistance both in Gram-negative and Gram-positive bacteria [93], sporadic cases of TG resistance have been reported recently. TG is not a substrate of the efflux pump *tet*(B), whose coding genes were identified in 19.6% of the isolates [94]. Nevertheless, a big fraction (74%) of our isolates were non-susceptible to TG. This far exceeds the numbers reported by other studies in the North African region [12,50]. While little is known about the mechanisms underlying TG resistance, few primarily chromosomal mechanisms have been confirmed for TG resistance in *A. baumannii*. These include target site alteration by mutations affecting the ribosomal protein S10 (*rpsJ*) [31,32] and overactive resistance nodulation division (RND) efflux pumps, particularly AdeABC [33], which, in turn, is controlled by the two-component systems *ade*SR [95] and BaeSR [34]. The first enzymatic resistance mechanism to TG was reported by [96], who described the TG-inactivating enzyme flavin-dependent monooxygenase *tet*(X). Plasmid-encoded *tet*(X), which confers high-level TG resistance, has been described previously [97,98]. Of the aforementioned mechanisms, only mutations affecting *adeS* (N268Y) and *baeS* (A436T) were exclusively found in TG-resistant isolates compared with TG-sensitive isolates of the same ST. Mutations affecting AdeABC regulatory systems were also reported as the sole TG-resistance mechanisms in *A. baumannii* in other studies [99]. While the amino acid alteration N268H was reported before as polymorphism in GC2 isolates [100], the impact of N268Y on the expression level of AdeABC and, hence, TG resistance has yet to be investigated. A recent study by Sawant et al. [101] reported the same prevalence of TG-resistant isolates as in our collection. The authors additionally demonstrated the overexpression of the *ade*ABC efflux pump associated with *adeRS* mutations including N268H. In addition to the previously mentioned TG resistance mechanisms, a recent study by Foong et al. [102] concluded that the MFS efflux pump *tet*A is up-regulated upon exposure to TG. The single-component pump then expels TG to the periplasm, from where it is then extruded by the tripartite RND efflux pumps *ade*ABC and *ade*IJK. The gene *tet*(A) was carried by only one isolate in our collection. However, the isolate was, surprisingly, susceptible to TG.

Of all our isolates, a subset of GC2 co-harbouring $bla_{\text{TEM-1}}$ and $bla_{\text{OXA-23}}$ can be considered to be of the highest public health concern. They carried the highest number of acquired resistance genes and exhibited resistance to the majority of antimicrobial classes. Moreover, they comprised 54.8% of GC2 and 36.9% of all isolates, which reflects their epidemic potential. In addition, GC5, GC4, and finally GC9 should also be closely monitored, due to their prevalence and the number of resistance genes. Although isolate 20Y0077 allocated to GC7 carried 13 acquired resistance genes, no reliable conclusion could be drawn about the threat imparted by GC7 as it was only represented by a single isolate. Additionally, the GC7-specific resistome could not be deduced from a single isolate. Further studies including a larger number of isolates belonging to all GCs are still required.

We also attempted to provide helpful markers for specific clones that might be utilised for inferring the burden of GCs in healthcare facilities with limited resources and where the seven locus-based MLST is not feasible. Such markers are based on antimicrobial resistance genes that might be routinely screened in microbiology laboratories. For this purpose, we categorised the antimicrobial resistance genes into three categories: unique, fully conserved, and not fully conserved. Of these, the unique genes that are fully conserved are best utilised for roughly assigning GCs. Concerning Table 3, unique genes mostly included *bla*$_{ADC}$ and *bla*$_{OXA-51-like}$ gene alleles. Similar findings were also reported by others [1,12,15,24,55,103]. Expanding our analysis to include the results of other studies revealed more intra-clonal conservation for *bla*$_{OXA-51}$ compared with *bla*$_{ADC}$ alleles [57,103]. While single-locus sequence typing might be required for defining specific *bla*$_{ADC}$ and *bla*$_{OXA-51-like}$ alleles, PCR detection of combinations of resistance genes fully conserved in each GC might provide an alternative epidemiologic tool within every laboratory's reach.

4. Materials and Methods

4.1. Bacterial Isolates

A collection of nonreplicated 60 *A. baumannii* isolates recovered from various clinical specimens and showing non-susceptibility to at least three classes of antimicrobial agents were selected for WGS. All isolates were obtained from the strain collection at the microbiology laboratory of the Faculty of Pharmacy, October University for Modern Sciences and Arts; they were originally recovered from patients admitted to tertiary care hospitals in Cairo, Egypt, in 2017 as part of routine clinical care. The specimens from which the isolates were recovered included blood (33; 55.0%), endotracheal tubes (10; 16.7%), sputum (7; 11.7%), wound swabs (6; 10.0%), catheter tips (3; 5.0%), and pus (1; 1.7%). All isolates were preliminarily identified using conventional microbiological techniques [104]. The identity of the isolates was subsequently confirmed using Matrix-Assisted Laser Desorption/Ionisation Time-of-Flight Mass Spectrometry (MALDI-TOF MS) and PCR detection of intrinsic *bla*$_{OXA-51-like}$ carbapenemase, *bla*$_{OXA-23-like}$, *bla*$_{OXA-24-like}$, *bla*$_{OXA-58-like}$, and IS*Aba1* genes as previously described [105].

4.2. Antimicrobial Susceptibility Testing

The initial screening for antimicrobial susceptibility was done using the Kirby–Bauer disk diffusion test performed and interpreted according to the Clinical and Laboratory Standards Institute (CLSI) criteria [106]. This was later confirmed by determining the minimum inhibitory concentrations (MICs) with broth microdilution assay in Mueller Hinton II broth using an automated MICRONAUT-S system (MICRONAUT, MERLIN Diagnostics GmbH, Bornheim-Hersel Germany). Micronaut-S-MDR-MRGN-Screening 3 and Micronaut-S β-Lactamases MIC plates were used for determining the MIC values of a large panel of antimicrobial agents according to the manufacturer's instructions [106]. The tested panel of antimicrobial agents included: penicillins (piperacillin and piperacillin/tazobactam); cephalosporins (cefotaxime, ceftazidime, ceftazidime/Avibactam, cefepime, and ceftolozane/tazobactam); carbapenems (imipenem, meropenem, and ertapenem); lipopeptides (colistin); aminoglycosides (amikacin); fluoroquinolones (ciprofloxacin and levofloxacin); and folate pathway antagonists (trimethoprim/sulfamethoxazole). In addition, susceptibility to tigecycline (TG), chloramphenicol, and fosfomycin was tested. According to the definitions proposed by Magiorakos et al. [107], the isolates were classified into multidrug-resistant (MDR), extensively drug-resistant (XDR), and pan-drug-resistant (PDR). Isolates that were non-susceptible to at least one agent in three or more antimicrobial classes were considered MDR, while the XDR phenotype was assigned to the isolates that were susceptible to only one or two antimicrobial classes. PDR was defined as non-susceptibility to all agents in all antimicrobial classes.

4.3. DNA Extraction, Library Preparation, and Whole-Genome Sequencing

Genomic DNA was extracted from pure colonies using a High Pure PCR Template Preparation Kit (Roche Diagnostics GmbH, Mannheim, Germany) according to the manufacturer's instructions. DNA was sequenced using Illumina MiSeq technology. For this purpose, the genomic libraries of the samples were prepared using a Nextera XT library preparation kit (Illumina Inc., San Diego, CA, USA). The samples were run on a MiSeq platform (Illumina Inc., San Diego, CA, USA) in paired-end mode as previously described [86].

4.4. Genome Assembly and Annotation

The raw read quality was assessed using FASTQC v0.11.7 (https://www.bioinformatics.babraham.ac.uk/projects/fastqc/, accessed on 1 July 2022) and de novo assemblies were generated with Shovill v.1.0.4 (https://github.com/tseemann/shovill, accessed on 1 July 2022) using SPAdes. Assembly statistics were assessed with QUAST v5.0.2 [108] and coding sequences were annotated using Prokka v.1.14.5 [109].

4.5. Multi-Locus Sequence Typing and Phylogenetic Analysis

For multi-locus sequence typing of the isolates, an in silico approach was chosen. Using the tool mlst v2.19.0 (https://github.com/tseemann/mlst) with the schemes Pasteur and Oxford deposited in PubMLST [110] (accessed on 6 July 2022), the strains were assigned to MLST sequence types (STs). GoeBURST analysis was used to group the STs of the isolates into clonal complexes (CCs) that were then allocated to the GCs described previously [1,18,20,26,111,112]. To do this, a minimum spanning tree combining all allelic profiles of both schemes was created using Phyloviz software v2.0 [113].

For a more in-depth strain differentiation, core genome single nucleotide polymorphisms (cgSNPs) were analysed. Here, data from foreign strains deposited in NCBI's Sequence Read Archive were also included (Table S1). SNP calling was carried out with Snippy v.4.6.0 (https://github.com/tseemann/snippy, accessed on 1 July 2022) using *A. baumannii* strain ATCC 17978 (GCF_001593425.2) as reference genome. The alignment of core genome SNPs was used as input for RAxML v8.2.12 [114] for Maximum likelihood analysis using the GTRGAMMA model of rate heterogeneity. The cgSNP-based phylogenetic tree was visualised using the interactive tree of life (iTOL) online tool v6.7 (https://itol.embl.de/itol.cgi, accessed on 12 March 2023).

4.6. Screening of Antimicrobial Resistance Determinants

Abricate v0.8.10 (https://github.com/tseemann/abricate, accessed on 1 July 2022) was used for screening the newly assembled isolates for potential antimicrobial resistance determinants. For this purpose, the databases Resfinder [115], CARD [116], and AMRFinder [117] were used. Mutations previously linked to reduced susceptibility to fluoroquinolones, colistin, and TG were investigated. With regard to the corresponding genes in *A. baumannii* ATCC 19606, mutations were analysed using the MSA and SNP analysis tool provided by the Bacterial and Viral Bioinformatics Resource Center (BV-BRC), available at: https://www.bv-brc.org/ (accessed on 1 March 2023). Multiple sequence alignments of the predicted amino acid sequences were done using the Clustal Omega (1.2.4) multiple sequence alignment service provided by the EMBL's European Bioinformatics Institute (EMBL-EBI) website (https://www.ebi.ac.uk/Tools/msa/clustalo/, accessed on 12 March 2023) and visualised in a coloured alignment format using the multiple alignment viewer tool MView 1.63, also hosted by EMBL-EBI and available at (https://www.ebi.ac.uk/Tools/msa/mview/, accessed on 12 March 2023).

4.7. Statistical Analysis

Chi-square and Fisher's Exact Tests were used for assessing the association between categorical variables such as resistance phenotypes, resistance determinants, GCs, and STs. All tests were two-tailed and were performed using IBM Statistical Package for the Social Sciences (SPSS) Statistics for Windows version 20.0 (IBM Corp., Armonk, NY, USA). p-values ≤ 0.05 were considered statistically significant.

5. Conclusions

Here, we demonstrated the contribution of five GCs in *A. baumannii* infections in Egypt, all showing MDR, XDR, or PDR phenotypes. We defined the resistome structure of the isolates and the potential contribution of specific resistance mechanisms. In more detail, we analysed the genetic background of non-susceptibility to last-line antimicrobials in silico. The association of different combinations of resistance mechanisms with resistance phenotypes was also examined. A preliminary image was drawn for the GC-specific resistomes.

Supplementary Materials: The following supporting information can be downloaded at: https://www.mdpi.com/article/10.3390/antibiotics12071149/s1, Table S1: SRA accessions and metadata of foreign strains included in the cgSNP-based phylogeny analysis; Table S2: Metadata and STs of the isolates included in the current study; Table S3: Post-assembly metrics of the draft genomes generated in the current study; Figure S1: A minimum spanning tree based on Pasteur MLST typing scheme of *A. baumannii*; Figure S2: A minimum spanning tree based on Oxford MLST typing scheme of *A. baumannii*; Figure S3: Multiple sequence alignment of the predicted amino acid sequences of CarO in all isolates compared with that of *A. baumannii* ATCC 19606; Figure S4: Multiple sequence alignment of the predicted amino acid sequences of AdeR in all isolates compared with that of *A. baumannii* ATCC 19606; Figure S5: Multiple sequence alignment of the predicted amino acid sequences of AdeS in all isolates compared with that of *A. baumannii* ATCC 19606; Figure S6: Multiple sequence alignment of the predicted amino acid sequences of BaeR in all isolates compared with that of *A. baumannii* ATCC 19606; Figure S7: Multiple sequence alignment of the predicted amino acid sequences of BaeS in all isolates compared with that of *A. baumannii* ATCC 19606; Table S4: Summary of mutations affecting *adeRS* and *baeRS* and correlation to tigecycline resistance.

Author Contributions: Conceptualisation and designing the experiments, G.W., W.F.E., L.D.S., M.W.P. and H.N.; sampling and identification, S.M.H.; bioinformatics analyses, H.B. and S.M.H.; data analysis, writing, review, and editing, S.M.H., W.F.E., H.B., A.S.G., S.M., D.F.K., L.D.S., M.W.P., G.W. and H.N. All authors have read and agreed to the published version of the manuscript.

Funding: This research was funded internally by the Friedrich-Loeffler-Institut and did not receive any external funding.

Institutional Review Board Statement: Ethical approval was not required for this study because no patient specimens or patient data were analyzed, only bacterial isolates stored in the microbiology laboratory of the Faculty of Pharmacy, October University for Modern Sciences and Arts were included.

Informed Consent Statement: Not applicable.

Data Availability Statement: The raw sequencing data were uploaded to the European Nucleotide Archive and are available under the project PRJEB58012.

Conflicts of Interest: The authors declare no conflict of interest.

References

1. Shelenkov, A.; Petrova, L.; Zamyatin, M.; Mikhaylova, Y.; Akimkin, V. Diversity of International High-Risk Clones of *Acinetobacter baumannii* Revealed in a Russian Multidisciplinary Medical Center during 2017–2019. *Antibiotics* **2021**, *10*, 1009. [CrossRef]
2. Harding, C.M.; Hennon, S.W.; Feldman, M.F. Uncovering the mechanisms of *Acinetobacter baumannii* virulence. *Nat. Rev. Microbiol.* **2018**, *16*, 91–102. [CrossRef] [PubMed]
3. Garnacho-Montero, J.; Timsit, J.F. Managing *Acinetobacter baumannii* infections. *Curr. Opin. Infect. Dis.* **2019**, *32*, 69–76. [CrossRef] [PubMed]

4. Elkhatib, W.F.; Khalil, M.A.F.; Ashour, H.M. Integrons and Antiseptic Resistance Genes Mediate Resistance of *Acinetobacter baumannii* and Pseudomonas aeruginosa Isolates from Intensive Care Unit Patients with Wound Infections. *Curr. Mol. Med.* **2019**, *19*, 286–293. [CrossRef]
5. Chen, C.T.; Wang, Y.C.; Kuo, S.C.; Shih, F.H.; Chen, T.L.; How, C.K.; Yang, Y.S.; Lee, Y.T. Community-acquired bloodstream infections caused by *Acinetobacter baumannii*: A matched case-control study. *J. Microbiol. Immunol. Infect.* **2018**, *51*, 629–635. [CrossRef]
6. Leung, W.S.; Chu, C.M.; Tsang, K.Y.; Lo, F.H.; Lo, K.F.; Ho, P.L. Fulminant community-acquired *Acinetobacter baumannii* pneumonia as a distinct clinical syndrome. *Chest* **2006**, *129*, 102–109. [CrossRef]
7. Karakonstantis, S.; Kritsotakis, E.I.; Gikas, A. Pandrug-resistant Gram-negative bacteria: A systematic review of current epidemiology, prognosis and treatment options. *J. Antimicrob. Chemother.* **2020**, *75*, 271–282. [CrossRef] [PubMed]
8. Gonzalez-Villoria, A.M.; Valverde-Garduno, V. Antibiotic-Resistant *Acinetobacter baumannii* Increasing Success Remains a Challenge as a Nosocomial Pathogen. *J. Pathog.* **2016**, *2016*, 7318075. [CrossRef] [PubMed]
9. Hsueh, P.R.; Teng, L.J.; Chen, C.Y.; Chen, W.H.; Yu, C.J.; Ho, S.W.; Luh, K.T. Pandrug-resistant *Acinetobacter baumannii* causing nosocomial infections in a university hospital, Taiwan. *Emerg. Infect. Dis.* **2002**, *8*, 827–832. [CrossRef]
10. Papathanakos, G.; Andrianopoulos, I.; Papathanasiou, A.; Priavali, E.; Koulenti, D.; Koulouras, V. Colistin-Resistant *Acinetobacter Baumannii* Bacteremia: A Serious Threat for Critically Ill Patients. *Microorganisms* **2020**, *8*, 287. [CrossRef]
11. WHO. *Global Priority List of Antibiotic-Resistant Bacteria to Guide Research, Discovery, and Development of New Antibiotics*; World Health Organization: Geneva, Switzerland, 2017.
12. Jalal, D.; Elzayat, M.G.; Diab, A.A.; El-Shqanqery, H.E.; Samir, O.; Bakry, U.; Hassan, R.; Elanany, M.; Shalaby, L.; Sayed, A.A. Deciphering Multidrug-Resistant *Acinetobacter baumannii* from a Pediatric Cancer Hospital in Egypt. *mSphere* **2021**, *6*, e0072521. [CrossRef] [PubMed]
13. Al-Agamy, M.H.; Khalaf, N.G.; Tawfick, M.M.; Shibl, A.M.; El Kholy, A. Molecular characterization of carbapenem-insensitive *Acinetobacter baumannii* in Egypt. *Int. J. Infect. Dis.* **2014**, *22*, 49–54. [CrossRef] [PubMed]
14. Elwakil, W.H.; Rizk, S.S.; El-Halawany, A.M.; Rateb, M.E.; Attia, A.S. Multidrug-Resistant *Acinetobacter baumannii* Infections in the United Kingdom versus Egypt: Trends and Potential Natural Products Solutions. *Antibiotics* **2023**, *12*, 77. [CrossRef] [PubMed]
15. Hamed, S.M.; Hussein, A.F.A.; Al-Agamy, M.H.; Radwan, H.H.; Zafer, M.M. Genetic Configuration of Genomic Resistance Islands in *Acinetobacter baumannii* Clinical Isolates From Egypt. *Front. Microbiol.* **2022**, *13*, 878912. [CrossRef] [PubMed]
16. Zafer, M.M.; Hussein, A.F.A.; Al-Agamy, M.H.; Radwan, H.H.; Hamed, S.M. Genomic Characterization of Extensively Drug-Resistant NDM-Producing *Acinetobacter baumannii* Clinical Isolates With the Emergence of Novel bla ADC-257. *Front. Microbiol.* **2021**, *12*, 736982. [CrossRef] [PubMed]
17. Hrabak, J.; Stolbova, M.; Studentova, V.; Fridrichova, M.; Chudackova, E.; Zemlickova, H. NDM-1 producing *Acinetobacter baumannii* isolated from a patient repatriated to the Czech Republic from Egypt, July 2011. *Eurosurveillance* **2012**, *17*, 20085. [CrossRef]
18. Zarrilli, R.; Pournaras, S.; Giannouli, M.; Tsakris, A. Global evolution of multidrug-resistant *Acinetobacter baumannii* clonal lineages. *Int. J. Antimicrob. Agents* **2013**, *41*, 11–19. [CrossRef]
19. Müller, C.; Stefanik, D.; Wille, J.; Hackel, M.; Higgins, P.G.; Siefert, H. Molecular epidemiology of carbapenem-resistant Acinetobacter baumannii clinical isolates and identification of the novel international clone IC9: Results from a worldwide surveillance study (2012–2016). In Proceedings of the ECCMID 2019: Proceeding of the 29th European Congress of Clinical Microbiology & Infectious Diseases, Amsterdam, The Netherlands, 13–16 April 2019.
20. Al-Hassan, L.; Elbadawi, H.; Osman, E.; Ali, S.; Elhag, K.; Cantillon, D.; Wille, J.; Seifert, H.; Higgins, P.G. Molecular Epidemiology of Carbapenem-Resistant *Acinetobacter baumannii* From Khartoum State, Sudan. *Front. Microbiol.* **2021**, *12*, 628736. [CrossRef]
21. Holt, K.; Kenyon, J.J.; Hamidian, M.; Schultz, M.B.; Pickard, D.J.; Dougan, G.; Hall, R. Five decades of genome evolution in the globally distributed, extensively antibiotic-resistant *Acinetobacter baumannii* global clone 1. *Microb. Genom.* **2016**, *2*, e000052. [CrossRef]
22. Diancourt, L.; Passet, V.; Nemec, A.; Dijkshoorn, L.; Brisse, S. The population structure of *Acinetobacter baumannii*: Expanding multiresistant clones from an ancestral susceptible genetic pool. *PLoS ONE* **2010**, *5*, e10034. [CrossRef]
23. Salloum, T.; Tannous, E.; Alousi, S.; Arabaghian, H.; Rafei, R.; Hamze, M.; Tokajian, S. Genomic mapping of ST85 blaNDM-1 and blaOXA-94 producing *Acinetobacter baumannii* isolates from Syrian Civil War Victims. *Int. J. Infect. Dis.* **2018**, *74*, 100–108. [CrossRef] [PubMed]
24. Fernandez-Cuenca, F.; Perez-Palacios, P.; Galan-Sanchez, F.; Lopez-Cerero, L.; Lopez-Hernandez, I.; Lopez Rojas, R.; Arca-Suarez, J.; Diaz-de Alba, P.; Rodriguez Iglesias, M.; Pascual, A. First identification of bla(NDM-1) carbapenemase in bla(OXA-94)-producing *Acinetobacter baumannii* ST85 in Spain. *Enferm. Infecc. Y Microbiol. Clin.* **2020**, *38*, 11–15. [CrossRef] [PubMed]
25. Karah, N.; Sundsfjord, A.; Towner, K.; Samuelsen, O. Insights into the global molecular epidemiology of carbapenem non-susceptible clones of *Acinetobacter baumannii. Drug Resist. Updates* **2012**, *15*, 237–247. [CrossRef]
26. Cerezales, M.; Xanthopoulou, K.; Wille, J.; Bustamante, Z.; Seifert, H.; Gallego, L.; Higgins, P.G. *Acinetobacter baumannii* analysis by core genome multi-locus sequence typing in two hospitals in Bolivia: Endemicity of international clone 7 isolates (CC25). *Int. J. Antimicrob. Agents* **2019**, *53*, 844–849. [CrossRef]

27. Zhang, G.; Leclercq, S.O.; Tian, J.; Wang, C.; Yahara, K.; Ai, G.; Liu, S.; Feng, J. A new subclass of intrinsic aminoglycoside nucleotidyltransferases, ANT(3″)-II, is horizontally transferred among *Acinetobacter* spp. by homologous recombination. *PLoS Genet.* **2017**, *13*, e1006602. [CrossRef]
28. Turton, J.F.; Woodford, N.; Glover, J.; Yarde, S.; Kaufmann, M.E.; Pitt, T.L. Identification of *Acinetobacter baumannii* by detection of the blaOXA-51-like carbapenemase gene intrinsic to this species. *J. Clin. Microbiol.* **2006**, *44*, 2974–2976. [CrossRef]
29. Peleg, A.Y.; Seifert, H.; Paterson, D.L. *Acinetobacter baumannii*: Emergence of a successful pathogen. *Clin. Microbiol. Rev.* **2008**, *21*, 538–582. [CrossRef]
30. Nigro, S.J.; Hall, R.M. Amikacin resistance plasmids in extensively antibiotic-resistant GC2 *Acinetobacter baumannii* from two Australian hospitals. *J. Antimicrob. Chemother.* **2014**, *69*, 3435–3437. [CrossRef]
31. Hammerstrom, T.G.; Beabout, K.; Clements, T.P.; Saxer, G.; Shamoo, Y. *Acinetobacter baumannii* Repeatedly Evolves a Hypermutator Phenotype in Response to Tigecycline That Effectively Surveys Evolutionary Trajectories to Resistance. *PLoS ONE* **2015**, *10*, e0140489. [CrossRef] [PubMed]
32. Beabout, K.; Hammerstrom, T.G.; Perez, A.M.; Magalhaes, B.F.; Prater, A.G.; Clements, T.P.; Arias, C.A.; Saxer, G.; Shamoo, Y. The ribosomal S10 protein is a general target for decreased tigecycline susceptibility. *Antimicrob. Agents Chemother.* **2015**, *59*, 5561–5566. [CrossRef]
33. Hornsey, M.; Ellington, M.J.; Doumith, M.; Thomas, C.P.; Gordon, N.C.; Wareham, D.W.; Quinn, J.; Lolans, K.; Livermore, D.M.; Woodford, N. AdeABC-mediated efflux and tigecycline MICs for epidemic clones of *Acinetobacter baumannii*. *J. Antimicrob. Chemother.* **2010**, *65*, 1589–1593. [CrossRef]
34. Lin, M.F.; Lin, Y.Y.; Yeh, H.W.; Lan, C.Y. Role of the BaeSR two-component system in the regulation of *Acinetobacter baumannii* adeAB genes and its correlation with tigecycline susceptibility. *BMC Microbiol.* **2014**, *14*, 119. [CrossRef] [PubMed]
35. Ghaith, D.M.; Zafer, M.M.; Al-Agamy, M.H.; Alyamani, E.J.; Booq, R.Y.; Almoazzamy, O. The emergence of a novel sequence type of MDR *Acinetobacter baumannii* from the intensive care unit of an Egyptian tertiary care hospital. *Ann. Clin. Microbiol. Antimicrob.* **2017**, *16*, 34. [CrossRef] [PubMed]
36. Sherif, M.M.; Elkhatib, W.F.; Khalaf, W.S.; Elleboudy, N.S.; Abdelaziz, N.A. Multidrug Resistant *Acinetobacter baumannii* Biofilms: Evaluation of Phenotypic-Genotypic Association and Susceptibility to Cinnamic and Gallic Acids. *Front. Microbiol.* **2021**, *12*, 716627. [CrossRef] [PubMed]
37. Attia, N.M.; Elbaradei, A. Fluoroquinolone resistance conferred by gyrA, parC mutations, and AbaQ efflux pump among *Acinetobacter baumannii* clinical isolates causing ventilator-associated pneumonia. *Acta Microbiològica Immunol. Hung.* **2019**, *67*, 234–238. [CrossRef]
38. Abdulzahra, A.T.; Khalil, M.A.F.; Elkhatib, W.F. First report of colistin resistance among carbapenem-resistant *Acinetobacter baumannii* isolates recovered from hospitalized patients in Egypt. *New Microbes New Infect.* **2018**, *26*, 53–58. [CrossRef] [PubMed]
39. Fam, N.S.; Gamal, D.; Mohamed, S.H.; Wasfy, R.M.; Soliman, M.S.; El-Kholy, A.A.; Higgins, P.G. Molecular Characterization of Carbapenem/Colistin-Resistant *Acinetobacter baumannii* Clinical Isolates from Egypt by Whole-Genome Sequencing. *Infect. Drug Resist.* **2020**, *13*, 4487–4493. [CrossRef]
40. Nepka, M.; Perivolioti, E.; Kraniotaki, E.; Politi, L.; Tsakris, A.; Pournaras, S. In Vitro Bactericidal Activity of Trimethoprim-Sulfamethoxazole Alone and in Combination with Colistin against Carbapenem-Resistant *Acinetobacter baumannii* Clinical Isolates. *Antimicrob. Agents Chemother.* **2016**, *60*, 6903–6906. [CrossRef]
41. Lee, C.R.; Lee, J.H.; Park, M.; Park, K.S.; Bae, I.K.; Kim, Y.B.; Cha, C.J.; Jeong, B.C.; Lee, S.H. Biology of *Acinetobacter baumannii*: Pathogenesis, Antibiotic Resistance Mechanisms, and Prospective Treatment Options. *Front. Cell. Infect. Microbiol.* **2017**, *7*, 55. [CrossRef]
42. Kim, D.H.; Choi, J.Y.; Kim, H.W.; Kim, S.H.; Chung, D.R.; Peck, K.R.; Thamlikitkul, V.; So, T.M.; Yasin, R.M.; Hsueh, P.R.; et al. Spread of carbapenem-resistant *Acinetobacter baumannii* global clone 2 in Asia and AbaR-type resistance islands. *Antimicrob. Agents Chemother.* **2013**, *57*, 5239–5246. [CrossRef]
43. Wareth, G.; Linde, J.; Nguyen, N.H.; Nguyen, T.N.M.; Sprague, L.D.; Pletz, M.W.; Neubauer, H. WGS-Based Analysis of Carbapenem-Resistant *Acinetobacter baumannii* in Vietnam and Molecular Characterization of Antimicrobial Determinants and MLST in Southeast Asia. *Antibiotics* **2021**, *10*, 563. [CrossRef]
44. Hujer, K.M.; Hamza, N.S.; Hujer, A.M.; Perez, F.; Helfand, M.S.; Bethel, C.R.; Thomson, J.M.; Anderson, V.E.; Barlow, M.; Rice, L.B.; et al. Identification of a new allelic variant of the *Acinetobacter baumannii* cephalosporinase, ADC-7 beta-lactamase: Defining a unique family of class C enzymes. *Antimicrob. Agents Chemother.* **2005**, *49*, 2941–2948. [CrossRef] [PubMed]
45. Lopes, B.S.; Al-Agamy, M.H.; Ismail, M.A.; Shibl, A.M.; Al-Qahtani, A.A.; Al-Ahdal, M.N.; Forbes, K.J. The transferability of blaOXA-23 gene in multidrug-resistant *Acinetobacter baumannii* isolates from Saudi Arabia and Egypt. *Int. J. Med. Microbiol.* **2015**, *305*, 581–588. [CrossRef] [PubMed]
46. El-Sayed-Ahmed, M.A.; Amin, M.A.; Tawakol, W.M.; Loucif, L.; Bakour, S.; Rolain, J.M. High prevalence of bla(NDM-1) carbapenemase-encoding gene and 16S rRNA armA methyltransferase gene among *Acinetobacter baumannii* clinical Isolates in Egypt. *Antimicrob. Agents Chemother.* **2015**, *59*, 3602–3605. [CrossRef]
47. Codjoe, F.S.; Donkor, E.S. Carbapenem Resistance: A Review. *Med. Sci.* **2017**, *6*, 1. [CrossRef] [PubMed]
48. Nicolau, D.P. Carbapenems: A potent class of antibiotics. *Expert Opin. Pharmacother.* **2008**, *9*, 23–37. [CrossRef]

49. Yoon, Y.K.; Yang, K.S.; Lee, S.E.; Kim, H.J.; Sohn, J.W.; Kim, M.J. Effects of Group 1 versus Group 2 carbapenems on the susceptibility of *Acinetobacter baumannii* to carbapenems: A before and after intervention study of carbapenem-use stewardship. *PLoS ONE* **2014**, *9*, e99101. [CrossRef]
50. Hassan, R.; Mukhtar, A.; Hasanin, A.; Ghaith, D. Role of insertion sequence Aba-1 and AdeS in reduced tigecycline susceptibility in MDR-*Acinetobacter baumannii* clinical isolates from Cairo, Egypt. *J. Chemother.* **2018**, *30*, 89–94. [CrossRef]
51. Leelasupasri, S.; Santimaleeworagun, W.; Jitwasinkul, T. Antimicrobial Susceptibility among Colistin, Sulbactam, and Fosfomycin and a Synergism Study of Colistin in Combination with Sulbactam or Fosfomycin against Clinical Isolates of Carbapenem-Resistant *Acinetobacter baumannii*. *J. Pathog.* **2018**, *2018*, 3893492. [CrossRef] [PubMed]
52. Wei, W.J.; Yang, H.F. Synergy against extensively drug-resistant *Acinetobacter baumannii* in vitro by two old antibiotics: Colistin and chloramphenicol. *Int. J. Antimicrob. Agents* **2017**, *49*, 321–326. [CrossRef]
53. Paranos, P.; Vourli, S.; Pournaras, S.; Meletiadis, J. Assessing Clinical Potential of Old Antibiotics against Severe Infections by Multi-Drug-Resistant Gram-Negative Bacteria Using In Silico Modelling. *Pharmaceuticals* **2022**, *15*, 1501. [CrossRef]
54. Abouelfetouh, A.; Mattock, J.; Turner, D.; Li, E.; Evans, B.A. Diversity of carbapenem-resistant *Acinetobacter baumannii* and bacteriophage-mediated spread of the Oxa23 carbapenemase. *Microb. Genom.* **2022**, *8*, 000752. [CrossRef] [PubMed]
55. Al-Hassan, L.; Zafer, M.M.; El-Mahallawy, H. Multiple sequence types responsible for healthcare-associated *Acinetobacter baumannii* dissemination in a single centre in Egypt. *BMC Infect. Dis.* **2019**, *19*, 829. [CrossRef] [PubMed]
56. Khuntayaporn, P.; Kanathum, P.; Houngsaitong, J.; Montakantikul, P.; Thirapanmethee, K.; Chomnawang, M.T. Predominance of international clone 2 multidrug-resistant *Acinetobacter baumannii* clinical isolates in Thailand: A nationwide study. *Ann. Clin. Microbiol. Antimicrob.* **2021**, *20*, 19. [CrossRef] [PubMed]
57. Zhang, X.; Li, F.; Awan, F.; Jiang, H.; Zeng, Z.; Lv, W. Molecular Epidemiology and Clone Transmission of Carbapenem-Resistant *Acinetobacter baumannii* in ICU Rooms. *Front. Cell. Infect. Microbiol.* **2021**, *11*, 633817. [CrossRef]
58. Levy-Blitchtein, S.; Roca, I.; Plasencia-Rebata, S.; Vicente-Taboada, W.; Velasquez-Pomar, J.; Munoz, L.; Moreno-Morales, J.; Pons, M.J.; Del Valle-Mendoza, J.; Vila, J. Emergence and spread of carbapenem-resistant *Acinetobacter baumannii* international clones II and III in Lima, Peru. *Emerg. Microbes Infect.* **2018**, *7*, 119. [CrossRef]
59. Mathlouthi, N.; El Salabi, A.A.; Ben Jomaa-Jemili, M.; Bakour, S.; Al-Bayssari, C.; Zorgani, A.A.; Kraiema, A.; Elahmer, O.; Okdah, L.; Rolain, J.M.; et al. Early detection of metallo-beta-lactamase NDM-1- and OXA-23 carbapenemase-producing *Acinetobacter baumannii* in Libyan hospitals. *Int. J. Antimicrob. Agents* **2016**, *48*, 46–50. [CrossRef]
60. Mathlouthi, N.; Ben Lamine, Y.; Somai, R.; Bouhalila-Besbes, S.; Bakour, S.; Rolain, J.M.; Chouchani, C. Incidence of OXA-23 and OXA-58 Carbapenemases Coexpressed in Clinical Isolates of *Acinetobacter baumannii* in Tunisia. *Microb. Drug Resist.* **2018**, *24*, 136–141. [CrossRef]
61. Ababneh, Q.; Al-Rousan, E.; Jaradat, Z. Fresh produce as a potential vehicle for transmission of *Acinetobacter baumannii*. *Int. J. Food Contam.* **2022**, *9*, 5. [CrossRef]
62. Jacobmeyer, L.; Semmler, T.; Stamm, I.; Ewers, C. Genomic Analysis of *Acinetobacter baumannii* Isolates Carrying OXA-23 and OXA-58 Genes from Animals Reveals ST1 and ST25 as Major Clonal Lineages. *Antibiotics* **2022**, *11*, 1045. [CrossRef]
63. Hamed, S.M.; Elkhatib, W.F.; El-Mahallawy, H.A.; Helmy, M.M.; Ashour, M.S.; Aboshanab, K.M.A. Multiple mechanisms contributing to ciprofloxacin resistance among Gram negative bacteria causing infections to cancer patients. *Sci. Rep.* **2018**, *8*, 12268. [CrossRef] [PubMed]
64. Kyriakidis, I.; Vasileiou, E.; Pana, Z.D.; Tragiannidis, A. *Acinetobacter baumannii* Antibiotic Resistance Mechanisms. *Pathogens* **2021**, *10*, 373. [CrossRef] [PubMed]
65. Lee, J.K.; Lee, Y.S.; Park, Y.K.; Kim, B.S. Mutations in the gyrA and parC genes in ciprofloxacin-resistant clinical isolates of *Acinetobacter baumannii* in Korea. *Microbiol. Immunol.* **2005**, *49*, 647–653. [CrossRef] [PubMed]
66. Mohammed, M.A.; Salim, M.T.A.; Anwer, B.E.; Aboshanab, K.M.; Aboulwafa, M.M. Impact of target site mutations and plasmid associated resistance genes acquisition on resistance of *Acinetobacter baumannii* to fluoroquinolones. *Sci. Rep.* **2021**, *11*, 20136. [CrossRef]
67. Roy, S.; Chatterjee, S.; Bhattacharjee, A.; Chattopadhyay, P.; Saha, B.; Dutta, S.; Basu, S. Overexpression of Efflux Pumps, Mutations in the Pumps' Regulators, Chromosomal Mutations, and AAC(6′)-Ib-cr Are Associated With Fluoroquinolone Resistance in Diverse Sequence Types of Neonatal Septicaemic *Acinetobacter baumannii*: A 7-Year Single Center Study. *Front. Microbiol.* **2021**, *12*, 602724. [CrossRef]
68. Coyne, S.; Courvalin, P.; Perichon, B. Efflux-mediated antibiotic resistance in *Acinetobacter* spp. *Antimicrob. Agents Chemother.* **2011**, *55*, 947–953. [CrossRef]
69. Tal-Jasper, R.; Katz, D.E.; Amrami, N.; Ravid, D.; Avivi, D.; Zaidenstein, R.; Lazarovitch, T.; Dadon, M.; Kaye, K.S.; Marchaim, D. Clinical and Epidemiological Significance of Carbapenem Resistance in *Acinetobacter baumannii* Infections. *Antimicrob. Agents Chemother.* **2016**, *60*, 3127–3131. [CrossRef]
70. Nguyen, M.; Joshi, S.G. Carbapenem resistance in *Acinetobacter baumannii*, and their importance in hospital-acquired infections: A scientific review. *J. Appl. Microbiol.* **2021**, *131*, 2715–2738. [CrossRef]
71. Lima, W.G.; Silva Alves, G.C.; Sanches, C.; Fernandes, S.O.A.; de Paiva, M.C. Carbapenem-resistant *Acinetobacter baumannii* in patients with burn injury: A systematic review and meta-analysis. *Burns* **2019**, *45*, 1495–1508. [CrossRef]
72. Donald, H.M.; Scaife, W.; Amyes, S.G.; Young, H.K. Sequence analysis of ARI-1, a novel OXA beta-lactamase, responsible for imipenem resistance in *Acinetobacter baumannii* 6B92. *Antimicrob. Agents Chemother.* **2000**, *44*, 196–199. [CrossRef]

73. Higgins, P.G.; Dammhayn, C.; Hackel, M.; Seifert, H. Global spread of carbapenem-resistant *Acinetobacter baumannii*. *J. Antimicrob. Chemother.* **2010**, *65*, 233–238. [CrossRef] [PubMed]
74. Hassan, R.M.; Salem, S.T.; Hassan, S.I.M.; Hegab, A.S.; Elkholy, Y.S. Molecular characterization of carbapenem-resistant *Acinetobacter baumannii* clinical isolates from Egyptian patients. *PLoS ONE* **2021**, *16*, e0251508. [CrossRef] [PubMed]
75. El Bannah, A.M.S.; Nawar, N.N.; Hassan, R.M.M.; Salem, S.T.B. Molecular Epidemiology of Carbapenem-Resistant *Acinetobacter baumannii* in a Tertiary Care Hospital in Egypt: Clonal Spread of blaOXA-23. *Microb. Drug Resist.* **2018**, *24*, 269–277. [CrossRef] [PubMed]
76. Ramadan, R.A.; Gebriel, M.G.; Kadry, H.M.; Mosallem, A. Carbapenem-resistant *Acinetobacter baumannii* and *Pseudomonas aeruginosa*: Characterization of carbapenemase genes and E-test evaluation of colistin-based combinations. *Infect. Drug Resist.* **2018**, *11*, 1261–1269. [CrossRef] [PubMed]
77. Lopes, B.S.; Amyes, S.G.B. Role of ISAba1 and ISAba125 in governing the expression of blaADC in clinically relevant *Acinetobacter baumannii* strains resistant to cephalosporins. *J. Med. Microbiol.* **2012**, *61 Pt 8*, 1103–1108. [CrossRef]
78. Zander, E.; Chmielarczyk, A.; Heczko, P.; Seifert, H.; Higgins, P.G. Conversion of OXA-66 into OXA-82 in clinical *Acinetobacter baumannii* isolates and association with altered carbapenem susceptibility. *J. Antimicrob. Chemother.* **2013**, *68*, 308–311. [CrossRef]
79. Jaidane, N.; Naas, T.; Oueslati, S.; Bernabeu, S.; Boujaafar, N.; Bouallegue, O.; Bonnin, R.A. Whole-genome sequencing of NDM-1-producing ST85 *Acinetobacter baumannii* isolates from Tunisia. *Int. J. Antimicrob. Agents* **2018**, *52*, 916–921. [CrossRef]
80. Charfi-Kessis, K.; Mansour, W.; Khalifa, A.B.H.; Mastouri, M.; Nordmann, P.; Aouni, M.; Poirel, L. Multidrug-resistant *Acinetobacter baumannii* strains carrying the bla(OxA-23) and the bla(GES-11) genes in a neonatology center in Tunisia. *Microb. Pathog.* **2014**, *74*, 20–24. [CrossRef] [PubMed]
81. Zhong, X.; Wu, X.; Schweppe, D.K.; Chavez, J.D.; Mathay, M.; Eng, J.K.; Keller, A.; Bruce, J.E. In Vivo Cross-Linking MS Reveals Conservation in OmpA Linkage to Different Classes of beta-Lactamase Enzymes. *J. Am. Soc. Mass Spectrom.* **2020**, *31*, 190–195. [CrossRef] [PubMed]
82. Wasfi, R.; Rasslan, F.; Hassan, S.S.; Ashour, H.M.; Abd El-Rahman, O.A. Co-Existence of Carbapenemase-Encoding Genes in *Acinetobacter baumannii* from Cancer Patients. *Infect. Dis. Ther.* **2021**, *10*, 291–305. [CrossRef]
83. Mussi, M.A.; Limansky, A.S.; Viale, A.M. Acquisition of resistance to carbapenems in multidrug-resistant clinical strains of *Acinetobacter baumannii*: Natural insertional inactivation of a gene encoding a member of a novel family of beta-barrel outer membrane proteins. *Antimicrob. Agents Chemother.* **2005**, *49*, 1432–1440. [CrossRef]
84. Pajand, O.; Rezaee, M.A.; Nahaei, M.R.; Mahdian, R.; Aghazadeh, M.; Soroush, M.H.; Tabrizi, M.S.; Hojabri, Z. Study of the carbapenem resistance mechanisms in clinical isolates of *Acinetobacter baumannii*: Comparison of burn and non-burn strains. *Burns* **2013**, *39*, 1414–1419. [CrossRef]
85. Abbasi, E.; Goudarzi, H.; Hashemi, A.; Chirani, A.S.; Ardebili, A.; Goudarzi, M.; Sharahi, J.Y.; Davoudabadi, S.; Talebi, G.; Bostanghadiri, N. Decreased carO gene expression and OXA-type carbapenemases among extensively drug-resistant *Acinetobacter baumannii* strains isolated from burn patients in Tehran, Iran. *Acta Microbiol. Immunol. Hung.* **2020**, *68*, 48–54. [CrossRef]
86. Wareth, G.; Linde, J.; Hammer, P.; Nguyen, N.H.; Nguyen, T.N.M.; Splettstoesser, W.D.; Makarewicz, O.; Neubauer, H.; Sprague, L.D.; Pletz, M.W. Phenotypic and WGS-derived antimicrobial resistance profiles of clinical and non-clinical *Acinetobacter baumannii* isolates from Germany and Vietnam. *Int. J. Antimicrob. Agents* **2020**, *56*, 106127. [CrossRef]
87. Yoon, E.J.; Goussard, S.; Touchon, M.; Krizova, L.; Cerqueira, G.; Murphy, C.; Lambert, T.; Grillot-Courvalin, C.; Nemec, A.; Courvalin, P. Origin in *Acinetobacter guillouiae* and dissemination of the aminoglycoside-modifying enzyme Aph(3′)-VI. *mBio* **2014**, *5*, e01972-14. [CrossRef] [PubMed]
88. Hamed, S.M.; Hussein, A.F.A.; Al-Agamy, M.H.; Radwan, H.H.; Zafer, M.M. Tn7382, a novel composite transposon harboring bla(NDM-1) and aphA6 in *Acinetobacter baumannii*. *J. Glob. Antimicrob. Resist.* **2022**, *30*, 414–417. [CrossRef]
89. Xanthopoulou, K.; Urrutikoetxea-Gutierrez, M.; Vidal-Garcia, M.; Diaz de Tuesta Del Arco, J.L.; Sanchez-Urtaza, S.; Wille, J.; Seifert, H.; Higgins, P.G.; Gallego, L. First Report of New Delhi Metallo-beta-Lactamase-6 (NDM-6) in a Clinical *Acinetobacter baumannii* Isolate From Northern Spain. *Front. Microbiol.* **2020**, *11*, 589253. [CrossRef]
90. Vazquez-Lopez, R.; Solano-Galvez, S.G.; Juarez Vignon-Whaley, J.J.; Abello Vaamonde, J.A.; Padro Alonzo, L.A.; Rivera Resendiz, A.; Muleiro Alvarez, M.; Vega Lopez, E.N.; Franyuti-Kelly, G.; Alvarez-Hernandez, D.A.; et al. *Acinetobacter baumannii* Resistance: A Real Challenge for Clinicians. *Antibiotics* **2020**, *9*, 205. [CrossRef] [PubMed]
91. Moffatt, J.H.; Harper, M.; Harrison, P.; Hale, J.D.; Vinogradov, E.; Seemann, T.; Henry, R.; Crane, B.; St Michael, F.; Cox, A.D.; et al. Colistin resistance in *Acinetobacter baumannii* is mediated by complete loss of lipopolysaccharide production. *Antimicrob. Agents Chemother.* **2010**, *54*, 4971–4977. [CrossRef] [PubMed]
92. Arroyo, L.A.; Herrera, C.M.; Fernandez, L.; Hankins, J.V.; Trent, M.S.; Hancock, R.E. The pmrCAB operon mediates polymyxin resistance in *Acinetobacter baumannii* ATCC 17978 and clinical isolates through phosphoethanolamine modification of lipid A. *Antimicrob. Agents Chemother.* **2011**, *55*, 3743–3751. [CrossRef] [PubMed]
93. Jenner, L.; Starosta, A.L.; Terry, D.S.; Mikolajka, A.; Filonava, L.; Yusupov, M.; Blanchard, S.C.; Wilson, D.N.; Yusupova, G. Structural basis for potent inhibitory activity of the antibiotic tigecycline during protein synthesis. *Proc. Natl. Acad. Sci. USA* **2013**, *110*, 3812–3816. [CrossRef]
94. Livermore, D.M. Tigecycline: What is it, and where should it be used? *J. Antimicrob. Chemother.* **2005**, *56*, 611–614. [CrossRef]
95. Xu, C.; Bilya, S.R.; Xu, W. adeABC efflux gene in *Acinetobacter baumannii*. *New Microbes New Infect.* **2019**, *30*, 100549. [CrossRef]

96. Moore, I.F.; Hughes, D.W.; Wright, G.D. Tigecycline is modified by the flavin-dependent monooxygenase TetX. *Biochemistry* **2005**, *44*, 11829–11835. [CrossRef] [PubMed]
97. He, T.; Wang, R.; Liu, D.; Walsh, T.R.; Zhang, R.; Lv, Y.; Ke, Y.; Ji, Q.; Wei, R.; Liu, Z.; et al. Emergence of plasmid-mediated high-level tigecycline resistance genes in animals and humans. *Nat. Microbiol.* **2019**, *4*, 1450–1456. [CrossRef] [PubMed]
98. Sun, J.; Chen, C.; Cui, C.Y.; Zhang, Y.; Liu, X.; Cui, Z.H.; Ma, X.Y.; Feng, Y.; Fang, L.X.; Lian, X.L.; et al. Plasmid-encoded tet(X) genes that confer high-level tigecycline resistance in *Escherichia coli*. *Nat. Microbiol.* **2019**, *4*, 1457–1464. [CrossRef]
99. Zheng, W.; Huang, Y.; Wu, W.; Zhu, J.; Zhang, T. Analysis of Efflux Pump System and Other Drug Resistance Related Gene Mutations in Tigecycline-Resistant *Acinetobacter baumannii*. *Comput. Math. Methods Med.* **2023**, *2023*, 8611542. [CrossRef]
100. Yoon, E.J.; Courvalin, P.; Grillot-Courvalin, C. RND-type efflux pumps in multidrug-resistant clinical isolates of *Acinetobacter baumannii*: Major role for AdeABC overexpression and AdeRS mutations. *Antimicrob. Agents Chemother.* **2013**, *57*, 2989–2995. [CrossRef]
101. Sawant, A.R.; Pagal, S.; Amar, A.K.; Panda, L.; Devi, C.S.; Shashikala, P.; Kanungo, R.; Prashanth, K. Coexistence of blaNDM-1, blaOXA-51, blaOXA-23, and armA in conjunction with novel mutations detected in RND efflux pump regulators in tigecycline resistant clinical isolates of *Acinetobacter baumannii*. *Pathog. Dis.* **2022**, *80*, ftac020. [CrossRef] [PubMed]
102. Foong, W.E.; Wilhelm, J.; Tam, H.K.; Pos, K.M. Tigecycline efflux in *Acinetobacter baumannii* is mediated by TetA in synergy with RND-type efflux transporters. *J. Antimicrob. Chemother.* **2020**, *75*, 1135–1139. [CrossRef] [PubMed]
103. Higgins, P.G.; Hagen, R.M.; Kreikemeyer, B.; Warnke, P.; Podbielski, A.; Frickmann, H.; Loderstadt, U. Molecular Epidemiology of Carbapenem-Resistant *Acinetobacter baumannii* Isolates from Northern Africa and the Middle East. *Antibiotics* **2021**, *10*, 291. [CrossRef]
104. Bouvet, P.; Grimont, P. Taxonomy of the genus *Acinetobacter* with the recognition of *Acinetobacter baumannii* sp. nov. *Acinetobacter haemolyticus* sp. nov. *Acinetobacter johnsonii* sp. nov. and *Acinetobacter junii* sp. nov. and emended descriptions of *Acinetobacter calcoaceticus* and *Acinetobacter Lwofii*. *Int. J. Syst. Evol. Microbiol.* **1986**, *36*, 228–240.
105. Wareth, G.; Linde, J.; Hammer, P.; Splettstoesser, W.D.; Pletz, M.W.; Neubauer, H.; Sprague, L.D. Molecular Characterization of German *Acinetobacter baumannii* Isolates and Multilocus Sequence Typing (MLST) Analysis Based on WGS Reveals Novel STs. *Pathogens* **2021**, *10*, 690. [CrossRef]
106. CLSI. *Performance Standards for Antimicrobial Susceptibility Testing*, 30th ed.; CLSI supplement M100; Clinical and Laboratory Standards Institute: Wayne, PA, USA, 2020.
107. Magiorakos, A.P.; Srinivasan, A.; Carey, R.B.; Carmeli, Y.; Falagas, M.E.; Giske, C.G.; Harbarth, S.; Hindler, J.F.; Kahlmeter, G.; Olsson-Liljequist, B.; et al. Multidrug-resistant, extensively drug-resistant and pandrug-resistant bacteria: An international expert proposal for interim standard definitions for acquired resistance. *Clin. Microbiol. Infect.* **2012**, *18*, 268–281. [CrossRef]
108. Gurevich, A.; Saveliev, V.; Vyahhi, N.; Tesler, G. QUAST: Quality assessment tool for genome assemblies. *Bioinformatics* **2013**, *29*, 1072–1075. [CrossRef]
109. Seemann, T. Prokka: Rapid prokaryotic genome annotation. *Bioinformatics* **2014**, *30*, 2068–2069. [CrossRef]
110. Jolley, K.A.; Bray, J.E.; Maiden, M.C.J. Open-access bacterial population genomics: BIGSdb software, the PubMLST.org website and their applications. *Wellcome Open Res.* **2018**, *3*, 124. [CrossRef] [PubMed]
111. Woodford, N.; Turton, J.F.; Livermore, D.M. Multiresistant Gram-negative bacteria: The role of high-risk clones in the dissemination of antibiotic resistance. *FEMS Microbiol. Rev.* **2011**, *35*, 736–755. [CrossRef] [PubMed]
112. Camargo, C.H.; Yamada, A.Y.; Nagamori, F.O.; de Souza, A.R.; Tiba-Casas, M.R.; de Moraes Franca, F.A.; Porto, M.; de Lima Garzon, M.L.; Higgins, P.; Madalosso, G.; et al. Clonal spread of ArmA- and OXA-23-coproducing *Acinetobacter baumannii* International Clone 2 in Brazil during the first wave of the COVID-19 pandemic. *J. Med. Microbiol.* **2022**, *71*, 001509. [CrossRef] [PubMed]
113. Ribeiro-Goncalves, B.; Francisco, A.P.; Vaz, C.; Ramirez, M.; Carrico, J.A. PHYLOViZ Online: Web-based tool for visualization, phylogenetic inference, analysis and sharing of minimum spanning trees. *Nucleic Acids Res.* **2016**, *44*, W246–W251. [CrossRef]
114. Stamatakis, A. RAxML version 8: A tool for phylogenetic analysis and post-analysis of large phylogenies. *Bioinformatics* **2014**, *30*, 1312–1313. [CrossRef] [PubMed]
115. Zankari, E.; Hasman, H.; Cosentino, S.; Vestergaard, M.; Rasmussen, S.; Lund, O.; Aarestrup, F.M.; Larsen, M.V. Identification of acquired antimicrobial resistance genes. *J. Antimicrob. Chemother.* **2012**, *67*, 2640–2644. [CrossRef] [PubMed]
116. Jia, B.; Raphenya, A.R.; Alcock, B.; Waglechner, N.; Guo, P.; Tsang, K.K.; Lago, B.A.; Dave, B.M.; Pereira, S.; Sharma, A.N.; et al. CARD 2017: Expansion and model-centric curation of the comprehensive antibiotic resistance database. *Nucleic Acids Res.* **2017**, *45*, D566–D573. [CrossRef] [PubMed]
117. Feldgarden, M.; Brover, V.; Haft, D.H.; Prasad, A.B.; Slotta, D.J.; Tolstoy, I.; Tyson, G.H.; Zhao, S.; Hsu, C.H.; McDermott, P.F.; et al. Validating the AMRFinder Tool and Resistance Gene Database by Using Antimicrobial Resistance Genotype-Phenotype Correlations in a Collection of Isolates. *Antimicrob. Agents Chemother.* **2019**, *63*, e00483-19. [CrossRef]

antibiotics

MDPI

Article

Genetic Determinants of *Acinetobacter baumannii* Serum-Associated Adaptive Efflux-Mediated Antibiotic Resistance

Mikaeel Young [1], Michaelle Chojnacki [1], Catlyn Blanchard [1], Xufeng Cao [2], William L. Johnson [1], Daniel Flaherty [2,3,4,*] and Paul M. Dunman [1,*]

1 Department of Microbiology and Immunology, University of Rochester Medical Center, Rochester, NY 14642, USA; mikaeel_young@baylor.edu (M.Y.); mchojnacki@trudeauinstitute.org (M.C.); wl.johnsonpubhealth@gmail.com (W.L.J.)
2 Department of Medicinal Chemistry and Molecular Pharmacology, Purdue University College of Pharmacy, Lafayette, IN 47907, USA
3 Purdue Institute for Drug Discovery, West Lafayette, IN 47907, USA
4 Purdue Institute of Inflammation, Immunology and Infectious Disease, West Lafayette, IN 47907, USA
* Correspondence: dflaher@purdue.edu (D.F.); paul_dunman@urmc.rochester.edu (P.M.D.); Tel.: +1-765-494-4761 (D.F.); +1-585-276-5700 (P.M.D.)

Abstract: *Acinetobacter baumannii* is a nosocomial pathogen of serious healthcare concern that is becoming increasingly difficult to treat due to antibiotic treatment failure. Recent studies have revealed that clinically defined antibiotic-susceptible strains upregulate the expression of a repertoire of putative drug efflux pumps during their growth under biologically relevant conditions, e.g., in human serum, resulting in efflux-associated resistance to physiologically achievable antibiotic levels within a patient. This phenomenon, termed Adaptive Efflux Mediated Resistance (AEMR), has been hypothesized to account for one mechanism by which antibiotic-susceptible *A. baumannii* fails to respond to antibiotic treatment. In the current study, we sought to identify genetic determinants that contribute to *A. baumannii* serum-associated AEMR by screening a transposon mutant library for members that display a loss of the AEMR phenotype. Results revealed that mutation of a putative pirin-like protein, YhaK, results in a loss of AEMR, a phenotype that could be complemented by a wild-type copy of the *yhaK* gene and was verified in a second strain background. Ethidium bromide efflux assays confirmed that the loss of AEMR phenotype due to pirin-like protein mutation correlated with reduced overarching efflux capacity. Further, flow cytometry and confocal microscopy measures of a fluorophore 7-(dimethylamino)-coumarin-4-acetic acid (DMACA)-tagged levofloxacin isomer, ofloxacin, further verified that YhaK mutation reduces AEMR-mediated antibiotic efflux. RNA-sequencing studies revealed that YhaK may be required for the expression of multiple efflux-associated systems, including MATE and ABC families of efflux pumps. Collectively, the data indicate that the *A. baumannii* YhaK pirin-like protein plays a role in modulating the organism's adaptive efflux-mediated resistance phenotype.

Keywords: *Acinetobacter baumannii*; adaptive efflux; antibiotic tolerance; serum

1. Importance

Intrinsic *A. baumannii* antibiotic resistance is predominantly attributable to mutations that lead to overexpression of one or more of the organism's 40 putative efflux pump systems [1]. In addition to intrinsic resistance, it is well recognized that bacteria elicit adaptive resistance mechanisms during infection that allow them to resist physiologically achievable antibiotic concentrations within a patient (reviewed in [2]). In that regard, laboratory-defined antibiotic-susceptible *A. baumannii* has the remarkable ability to induce an expansive repertoire of drug efflux pumps during its growth under physiological salt conditions, e.g., in pulmonary surfactant or human serum, thereby allowing them to become resistant to physiologically achievable antibiotic concentrations within a patient [3–7]. This phenomenon, termed adaptive efflux-mediated antibiotic resistance (AEMR), is conserved across *A. baumannii* lineages as well as other bacterial species, such as *Pseudomonas aeruginosa* [8]. The current study was designed to define genetic determinants that modulate *A. baumannii* AEMR, as such factors may represent targets for the therapeutic development of agents that reduce antibiotic resistance.

2. Introduction

Antibiotic-resistant bacteria accounted for 2.8 million infections and 35,000 deaths in the U.S. in 2019 and are predicted to cause 10 million infections and 700,000 deaths globally by the year 2050 [9,10]. In that regard, *A. baumannii* has emerged as a particularly problematic pathogen. Indeed, the latest National Healthcare Safety Network (NHSN) report revealed that *A. baumannii* accounted for approximately 3% of all nosocomial infections, 7.8% of ventilator-associated pneumonias, and 2.2% of all hospital central line-associated bloodstream infections in the U.S. in 2006 [11]. Simultaneously *A. baumannii* antibiotic resistance has escalated at a rapid pace; 47.7% of clinical isolates collected from hospitalized patients in the U.S. between 2013 to 2017 were found to be multidrug-resistant, and a staggering 37.5% were not susceptible to carbapenems, which have been considered front-line therapeutics for severe *A. baumannii* infections [12]. In response, the World Health Organization, Centers for Disease Control and Prevention, and Infectious Diseases Society of America have all recognized *A. baumannii* as a pathogen of urgent healthcare concern, and the organism has been designated as one of the six ESKAPE (*Enterococcus faecium*, *Staphylococcus aureus*, *Klebsiella pneumoniae*, *A. baumannii*, *P. aeruginosa*, and *Enterobacter* sp.) bacterial pathogens of greatest concern that can 'escape' the effects of antibiotics due to resistance [9,13,14].

A. baumannii antibiotic resistance is, in part, attributable to the presence of antibiotic-modifying enzymes but is generally regarded to be a consequence of the organism's expansive repertoire of antibiotic efflux pumps. *A. baumannii* is capable of producing each of the six major classes of drug efflux pumps, including the resistance nodulation division (RND) pumps AdeABC, AdeFGH, and AdeIJK, the major facilitator superfamily (MFS) pumps CraA and AmvA, the multidrug and toxic compound extrusion (MATE) pump AbeM, small multidrug resistance (SMR) AbeS pump, proteobacterial antimicrobial compound efflux (PACE) pump, AceI, and the ATP-binding cassette (ABC) family of pump MacB, which collectively confer resistance to virtually all front-line antibiotics (reviewed in [1]). While a multitude of *A. baumannii* antibiotic efflux pumps have been characterized, comparative genomics indicates that the genomes of strains ATCC 17978 and/or AYE contain more than 45 putative additional drug efflux-like transporters, any one of which may also contribute to antibiotic resistance [8,15].

Antibiotic drug efflux pumps are tightly regulated, and most are repressed or only basally expressed during conventional laboratory growth conditions, requiring mutations that lead to their constitutive expression to achieve an intrinsic antibiotic resistance phenotype, but often at a significant fitness cost [16]. It is becoming increasingly appreciated that while such efflux pumps may not be expressed at a high level in ordinary laboratory conditions, they can be induced in response to stress conditions that an organism encounters during infection. Such induction may allow a seemingly laboratory-defined antibiotic-susceptible bacterial strain to tolerate or resist antibiotic treatment. Indeed, while *Escherichia coli* strain K12 codes for at least 36 efflux pumps, only one, AcrAB-TolC, is appreciably expressed during exponential growth in nutrient-rich media at 37 °C [17,18]. Conversely, under starvation conditions, *E. coli* induces at least 10 efflux systems that modulate resistance to the antibiotic ampicillin, whereas anaerobic growth and low pH induce expression of the RND-like efflux pump, MdtEF, and lead to antibiotic resistance [18–20]. Conditions that mimic the infection setting have also been shown to induce efflux systems in other bacterial species. For instance, *Mycobacterium tuberculosis*, *Bacillus cereus*, and *Salmonella enterica* induce efflux pumps that modulate antibiotic resistance in response to macrophage internalization, oxidative stress, and bile exposure, respectively [21–23].

Similarly, recent studies have revealed that laboratory-defined antibiotic-susceptible *A. baumannii* strains upregulate a multitude of efflux pumps in response to biologically relevant host conditions, such as physiological salt concentrations and human serum, and display efflux-mediated resistance to physiologically achievable concentrations of aminoglycosides, carbapenems, fluoroquinolones, tetracyclines, and the glycylcycline antibiotic, tigecycline, within a patient [7,8]. This phenomenon, termed adaptive efflux-mediated resistance (AEMR), has been hypothesized to account for one mechanism by which antibiotic-susceptible strains-resist antibiotic treatment within the host setting, ultimately leading to clinical antibiotic failure [24]. While serum-associated AEMR has been observed in *A. baumannii* as well as *P. aeruginosa* clinical isolates, the underlying genetic determinants that modulate this phenomenon are poorly understood [8,25]. Accordingly, the objective of the current study was to identify genetic determinants that contribute to *A. baumannii* AEMR. To do so, an *A. baumannii* transposon mutant library was screened for members that display a loss of serum-associated efflux-mediated antibiotic resistance phenotype. Complementation and efflux assays coupled with confocal microscopy, flow cytometry, and RNA sequencing (RNA-seq) analyses collectively revealed that the *A. baumannii* YhaK pirin-like protein (ATCC 17978 A1S_3277) is likely to be a key component of the organism's AEMR to the antibiotics minocycline and levofloxacin.

3. Materials and Methods

3.1. Bacterial Strains and Growth Conditions

The bacterial strains and plasmids used in this study are listed in Table 1. *A. baumannii* strain 98-37-09 is a well-characterized antibiotic-susceptible cerebrospinal fluid clinical isolate obtained from the Centers for Disease Control and Prevention [8,26–28]. The *A. baumannii* strain 98-37-09 EZ-Tn5 transposon mutant library used in these studies is comprised of 6000 members that have been individually arrayed in a 96-well plate format and have been previously described [28]. *A. baumannii* strain 98-37-09-YhaK is an EZ-Tn5-mutant library member that harbors a transposon insertion within the strain's YhaK gene (A1S_3277:Tn5), whereas strains 98-37-09-YhaK pWH1266 and 98-37-09-YhaK pYhaK are corresponding mutants that have been transformed with the plasmid shuttle-vector pWH1266 [29] or pWH1266 harboring a wild-type copy of the YhaK gene (pYhaK), respectively, and were created for this study as described below. *A. baumannii* strain AB5075-UW is a clinical isolate obtained from a combat wound infection, whereas strain AB5075-YhaK is an AB5075-UW derivative containing a transposon insertion with the YhaK locus (ABUW_0207:T26) and were obtained from the Manoil *A. baumannii* mutant library [30] from the University of Washington (Seattle, WA, USA). *A. baumannii* strains AB5075-YhaK pWH1266 and AB5075-YhaK pYhaK are AB5075-YhaK derivatives containing plasmid pWH1266 [29] and pYhaK, respectively, and created for this study, as described below. Chemically competent OneShot® TOP10 *Escherichia coli* were used for cloning and plasmid construction (Thermo Fisher Scientific, Waltham, MA, USA). All *A. baumannii* and *E. coli* strains were grown in Luria-Bertani (LB) broth and/or 100% human serum (Corning Life Sciences, Tewksbury, MA, USA) at 37 °C in media supplemented with 50 μg mL^{-1} of kanamycin or 10 μg mL^{-1} tetracycline, as indicated.

Table 1. Strains and plasmids used in this study.

Strains and Plasmids	Relevant Genotype/Phenotype	Source
A. baumannnii		
98-37-09	Wild type	[28]
98-37-09 YhaK	98-37-09 *yhaK*::EZTn5	[28]
98-37-09 YhaK pWH1266	98-37-09 *yhaK*::EZTn5; pWH1266	This study
98-37-09 YhaK pYhaK	98-37-09 *yhaK*::EZTn5; pYhaK	This study
AB5075-UW	Wild type	[30]
AB5075 YhaK	ABUW_0207-119::T26	[30]
AB5075 YhaK pWH1266	ABUW_0207-119::T26; pWH1266	This study
AB5075 YhaK pYhaK	ABUW_0207-119::T26; pYhaK	This study
E. coli		
OneShot® TOP10	F-*mcrA* Δ(*mrr-hsd*RMS-*mcr*BC) Φ80*lac*ZΔM15 Δ*lac*X74 *rec*A1 *ara*D139 Δ(*araleu*)7697 *gal*U *gal*K *rps*L (StrR) *end*A1 *nup*G	Thermo Fisher
Plasmids		
pWH1266	pBR322 derivative; tet^R and amp^R	[29]
pYhaK	pWH1266 containing 98-37-09 YhaK	This study

3.2. *Transposon Mutant Screening for Loss of Adaptive Efflux-Associated Resistance*

Each member of an *A. baumannii* strain 98-37-09 transposon mutant library was individually grown overnight in independent wells of a microtiter plate containing 100 μL LB broth supplemented with kanamycin at 37 °C. For screening, mutants were transferred to individual wells of a new microtiter plate (diluted 1:100) containing 100 μL fresh medium and grown to optical density 600_{nm} ($OD600_{nm}$) of approximately 0.4–0.5 at 37 °C. Approximately 1×10^5 colony forming units (CFU) of each transposon mutant library member were then transferred to individual wells of a 96-well round bottom plate containing 100 μL of 100% human serum supplemented with 1.0 μg mL^{-1} minocycline (0.5× MIC in human serum). Plates were incubated at 37 °C for 48 h, and bacterial growth was detected by the unaided human eye. Transposon library members that displayed a loss of growth phenotype were expected to include mutants that exhibited a loss of efflux-associated antibiotic tolerance and were selected for further study.

3.3. *Identification of Transposon Insertion Site of Mutants of Interest*

Inverse PCR was used to identify the transposon insertion sites of transposon library members of interest, as previously described [28]. Briefly, total bacterial DNA was purified from each mutant using DNeasy Blood and Tissue kits (Qiagen, Valencia, CA, USA). Two micrograms of DNA was digested with the restriction enzyme *Afe*I (10 U; New England Biolabs, Ipswich, MA, USA) for 2 h at 37 °C and heat-inactivated at 65 °C for 5 min. Restriction fragments were circularized by ligation using 1.5 units of T4 DNA ligase (Thermo Fisher, Waltham, MA, USA) for 16 h at 16 °C. Following ligase heat-inactivation, PCR was performed using Platinum PCR Supermix High Fidelity Kits (Invitrogen, Waltham, MA, USA) and transposon-specific primers (forward, 5′-ACCTACAACAAAGCTCTCATCAACC-3′; reverse, 5′-CTACCCTGTGGAACACCTACATCT-3′) in a GeneAmp PCR System 9700 thermocycler (Applied Biosystems, Foster City, CA, USA) with the following parameters: 94 °C for 10 min and 50 cycles of 94 °C for 30 s, 57 °C for 30 s, and 72 °C for 6 min, followed by an extension at 72 °C for 10 min. PCR products were electrophoresed in a 1% UltraPure Agarose gel (Invitrogen) at 75 V for 40 min and gel-purified using QIAquick Gel Extraction Kits (Qiagen). Purified products were cloned into plasmid pCR2.1-TOPO using TOPO TA cloning kits and then transformed into 50 μL TOP10 *E. coli*, following the manufacturer's recommendations (Thermo Fisher, Waltham, MA, USA). Transformants were selected on LB agar supplemented with 50 μg mL^{-1} kanamycin and screened for plasmid inserts by colony PCR. To do so, approximately half of each colony was combined with DreamTaq Green PCR Mastermix (Thermo Fisher, Waltham, MA, USA) and amplified using plasmid-directed M13 sequencing primers (Forward: 5′-GTAAAACGACGGCCAG-3′, Reverse 5′-CAGGAAACAGCTATGAC-3′) using the following parameters: 95 °C for 5 min and 35 cycles of 95 °C for 1 min, 54 °C for 1.5 min, and 72 °C for 5 min, followed by a final extension at 72 °C for 5 min. PCR products were electrophoresed in a 1% UltraPure Agarose gel. Colonies containing a plasmid insert were propagated, and plasmid DNA was isolated using the Qiaprep Spin Miniprep kit (Qiagen) and then sequenced by ACGT Incorporated (Wheeling, IL, USA) using M13 sequencing primers. Sequences were aligned utilizing BLAST against the *A. baumannii* strain 17978 (NCBI:txid400667) genome to map the transposon insertion site [31].

3.4. Adaptive Antibiotic Tolerance Assays

A. baumannii strains 98-37-09, AB5075, and corresponding mutants of interest were propagated overnight in LB broth at 37 °C, diluted into fresh medium (1:100), and grown to $OD600_{nm}$ 0.4–0.5 at 37 °C with aeration. A total of approximately 1×10^5 cells were transferred to individual wells of a 96-well round bottom plate (Corning Life Sciences, Corning, NY, USA) containing 100 µL of LB (negative control) or 100% human serum (AEMR inducing condition) supplemented with 0, 0.125, 0.25, 0.5, 1, 2, or 4.0 µg mL^{-1} minocycline, levofloxacin, ofloxacin, or with 0, 25, 50, 100, or 200 µg mL^{-1} DMACA-labeled ofloxacin (described below) and incubated at 37 °C for 48 h. To quantify the antimicrobial effects of each antibiotic, cells were serially diluted in 0.8% sodium chloride and plated on LB agar to enumerate the viable number of CFU mL^{-1} for each strain during growth in LB and serum supplemented with various concentrations of the indicated antibiotic. All experiments were performed at least 3 times, and results were averaged and presented with ±standard deviation.

3.5. Ethidium Bromide Efflux Assays

Bacterial ethidium bromide (EtBr) efflux assays were performed to assess the ability of the indicated *A. baumannii* strain to efflux EtBr, as previously described [26]. *A. baumannii* strains were propagated for 16 h in LB broth at 37 °C with aeration, diluted 1:100 in fresh LB (negative control) or 100% human serum, and grown to the mid-exponential phase. Cells were collected by centrifugation at 900× *g* for 20 min at 4 °C, washed three times with sodium phosphate buffer (15 mM Na_2HPO_4 and 5 mM NaH_2PO_4), and resuspended to a final $OD600_{nm}$ of 0.2 in sodium phosphate buffer. A total of 1×10^6 CFUs were transferred to individual wells of a 96-well white plate (Falcon, Corning Life Sciences), containing either dimethyl sulfoxide (DMSO; negative control) or 40 µg mL^{-1} of the efflux pump inhibitor phenylalanine arginine β-naphthylamide (PAβN; MP Biomedicals, Irvine, CA, USA). Ethidium bromide (10 µg mL^{-1}) was added, and corresponding fluorescence was monitored (excitation, 530_{nm}; emission, 600_{nm}) every 5 min for a total of 90 min on a SPECTRAmax5 fluorimeter (Molecular Devices, Sunnyvale, CA, USA) and averaged. All experiments were performed at least 3 times, and results were averaged and presented with ± standard deviation.

3.6. Outer Membrane Integrity Assay

Bacterial cell membrane permeability measures were performed using 1-*N*-phenylnapthylamine (NPN) assays in triplicate, as previously described [32]. Briefly, *A. baumannii* strains were propagated for 16 h in LB broth at 37 °C with aeration, diluted 1:50 in fresh human serum, and grown to the mid-exponential phase. Cells were harvested by centrifugation (3000× *g*), washed twice with assay buffer (5 mM HEPES, 5 mM glucose, pH 7.2), and resuspended in assay buffer to a final $OD600_{nm}$ = 1.0. In triplicate, a total of 100 µL of washed cells were mixed with an equal volume of assay buffer containing 20 µM NPN and placed into individual wells of a 96-well black plate (Corning Life Sciences, Corning, NY, USA); A volume of 5 µL of assay buffer or 0.06% Triton X-100 served as negative and positive controls, respectively. Fluorescence was measured at an excitation wavelength of 350_{nm}

and emission wavelength of 420$_{nm}$, averaged, and normalized to wild-type cells. All experiments were performed at least 3 times, and results were averaged and presented with ±standard deviation.

3.7. Construction of Complementation Plasmids

The *A. baumannii* 17978 YhaK locus A1S_3277 and 500 base pair upstream sequence were cloned from strain 98-37-09 and placed into the multiple cloning site of the shuttle vector pWH1266 [29]. To do so, 98-37-09 genomic DNA served as a template for PCR amplification using A1S_3277 forward primer 5′-ATGCCTGCAGAGTTGTTTTTACCAGTTGCTTTA ACCGC-3′ and reverse primer 5′-ATGCCTGCAGTTATCCTTCTTGCATGATTGAACCGAA TTTTCC-3′, with each oligonucleotide containing a PstI restriction enzyme site (underlined) using DreamTaq PCR Mastermix Kits (Thermo Fisher). Resulting PCR products were separated by 0.8% agarose gel electrophoresis, purified using QIAquick Gel Extraction Kits, ligated into plasmid pCR2.1-TOPO using TOPO TA cloning kits, and transformed into *E. coli* TOP10 cells following the manufacturer's recommendations (Thermo Fisher, Waltham, MA, USA). Transformants were selected by growth on LB agar supplemented with 50 μg mL^{-1} tetracycline and propagated, and plasmid DNA was isolated using Qiaprep Spin Miniprep kits (Qiagen). Purified plasmids were digested with PstI to liberate the plasmid insert, which was gel-purified using QIAquick Gel Extraction Kits and ligated into PstI-digested pWH1266 shuttle vector using T4 DNA ligase (Thermo Fisher) to generate plasmid pYhaK (A1S_3277). Constructs were first transformed into *E. coli* TOP10 for plasmid propagation, purified using Qiaprep Spin Miniprep kits, and sequenced to ensure their integrity (ACGT Incorporated, Wheeling, IL, USA). Plasmids pWH1266 and pYhaK were electroporated into *A. baumannii* 98-37-09-YhaK and AB5075-YhaK and selected with 50 μg mL^{-1} tetracycline for complementation studies.

3.8. Real-Time Quantitative PCR to Measure A1S_3277 Expression

In triplicate, *A. baumannii* strains were grown to exponential phase in either LB media or human serum in the absence or presence of 0.15 μg mL^{-1} levofloxacin. Total bacterial RNA was isolated using Qiagen RNeasy kits, DNase treated, repurified and used as a template (400 ng) for qRT-PCR using primers for A1S_3277 (forward, 5′-TCGATGAAGAAATTTCTTGGTGCATATCAAAATAACC-3′; reverse, 5′-ATGCCTGCAGT TATCCTTC TTGCATGATTGAACCGAATTTTCC-3′) and as a template (4 ng) for 16S rRNA (forward,5′-CTGTAGCGGGTCTGAGAGGAT-3′; reverse, 5′-CCATAAGCCTTCTTCACAC-3′) amplified, and measured using PerfeCta SYBR Green Fastmix and qScript cDNA SuperMix kits following the manufacturer's recommendations (QuantaBio, Beverly, MA, USA). All samples were conducted in triplicate and normalized to 16S rRNA, averaged, and compared to untreated exponential phase wild-type cells grown in LB media.

3.9. Synthesis of DMACA-Ofloxacin

In the first step, 9-Fluoro-3-methyl-7-oxo-10-(piperazin-1-yl)-2,3-dihydro-7*H*-[1,4] oxazino[2,3,4-*ij*]quinoline-6-carboxylic acid (Intermediate 1) was synthesized as follows. Under argon atmosphere, piperazine (1.03 g, 12.0 mmol, 3 eq) was added to a suspension of 9,10-difluoro-3-methyl-7-oxo-2,3-dihydro-7*H*-[1,4]oxazino[2,3,4-*ij*]quinoline-6-carboxylic acid (0.56 g, 2.00 mmol, 1 eq) in anhydrous DMSO (10 mL). The suspension was stirred at 95 °C for 16 h. Cold acetone was then added to the mixture, resulting brown precipitate that was triturated in acetone to give intermediate 1 (0.60 g, 1.74 mmol, yield 87%) as a pale-yellow solid. 1H NMR (500 MHz, DMSO-d6) δ 8.93 (s, 1H), 7.53 (d, *J* = 12.4 Hz, 1H), 4.88 (dd, *J* = 13.1, 6.4 Hz, 1H), 4.54 (d, *J* = 11.4 Hz, 1H), 4.38–4.25 (m, 1H), 2.69 (s, 4H), 2.47–2.45 (m, 4H), 1.40 (d, *J* = 6.7 Hz, 3H). ^{13}C NMR (126 MHz, DMSO-d_6) δ 176.63, 166.46, 156.82, 154.86, 146.48, 140.47, 140.42, 132.70, 125.14, 119.94, 119.86, 109.23, 107.10, 103.72, 103.53, 68.32, 55.10, 51.83, 46.50, 45.45, 18.25. MS (ESI, Advion Mass Spec LC MS) m/z: 348.1 (M+1)$^+$. In the second step, 10-(4-(2-(7-(dimethylamino)-2-oxo-2H-chromen-4-yl)acetyl)piperazin-1-yl)-9-fluoro-3-methyl-7-oxo-2,3 dihy-dro-7*H*-[1,4]oxazino[2,3,4-*ij*]quinoline-6-carboxylic acid (DMACA-Oflox) was created, as follows. Under an argon atmosphere, a solution was prepared of 2-(7-(dimethylamino)-2-oxo-2*H*-chromen-4-yl)acetic acid (DMACA, 0.22 g, 0.90 mmol, 1.5 eq), HOBt (0.18 g, 1.20 mmol, 2 eq), and (3-dimethylaminopropyl)-ethyl-carbodiimide HCl (0.23 g, 1.20 mmol, 2 eq) in *anhydrous N,N*-dimethylformamide (DMF, 8 mL) was stirred in an ice bath, then *N,N*-diisopropylethylamine (0.36 mL, 2.40 mmol) was added into the reaction mixture and then stirred at 0 °C for 30 min. A solution of intermediate 1 (0.21 g, 0.60 mmol, 1 eq) in anhydrous DMF (4 mL) was added, and the mixture was heated at 80 °C overnight. The reaction mixture was extracted with ethyl acetate and H_2O, the organic phase was washed with H_2O, Brine, then dried with anhydrous Na_2SO_4, filtered, and concentrated under vacuum, and the crude product was purified by manual silica gel column Chromatography (0% MeOH/DCM~10% MeOH/DCM) to afford DMACA-Oflox (0.030 g, yield 9%) as a pale yellow solid. ^{1}H NMR (500 MHz, $CDCl_3$) δ 8.63 (s, 1H), 7.75 (d, J = 11.8 Hz, 1H), 7.47 (d, J = 9.0 Hz, 1H), 6.64 (dd, J = 9.0, 2.4 Hz, 1H), 6.52 (d, J = 2.4 Hz, 1H), 6.02 (s, 1H), 4.49 (d, J = 11.3 Hz, 2H), 4.40–4.33 (m, 1H), 3.84 (s, 4H), 3.60 (s, 2H), 3.35 (t, J = 12.5 Hz, 4H), 3.07 (s, 6H), 1.62 (d, J = 6.7 Hz, 3H). MS (ESI, Advion Mass Spec LC-MS) m/z: 576.7 $(M+1)^+$.

3.10. Flow Cytometry

Flow cytometry was used to measure the cellular accumulation of the levofloxacin isomer, ofloxacin, labeled with the fluorophore 7-(dimethylamino)-coumarin-4-acetic acid (DMACA-Oflox; optimal excitation 390_{nm}/emission 480_{nm}; MIC 100 μM) in *A. baumannii* strains 98-37-09 and 98-37-09 YhaK cultured in LB or human serum; all conditions were repeated at least three times. To do so, strains were grown overnight in LB and then sub-cultured (1:50 dilution) in either LB or serum for 2.5 h at 37 °C, washed twice in PBS, and then resuspended in PBS to an $OD600_{nm}$ of 0.4. Approximately 2×10^7 cells were returned to 2 mL of fresh LB or serum-supplemented with 75 μM DMACA-Oflox in the presence or absence of 20 μg mL^{-1} *A. baumannii* efflux-pump inhibitor 4-[(4-Methoxybenzyl)oxy] benzene sulfonamide (AE0003; ref. [33]) and incubated for 30 min at 37 °C. Cells were washed and resuspended in PBS and flowed through a BD LSR II flow cytometer (BD Biosciences, Franklin Lakes, NJ, USA). Cells were excited with a violet 405_{nm} laser using a 450/50 bandpass filter. For each sample, fluorescence intensity and side-scatter measures were recorded for a total of 100,000 events and analyzed using Flow Cytometry Standard (FCS) Express (De Novo Software, Pasadena, CA, USA).

3.11. Confocal Microscopy

Confocal microscopy was used to measure the cellular association of DMACA-Oflox with *A. baumannii* strains 98-37-09 and 98-37-09 YhaK when cultured in LB or human serum. To do so, strains were grown overnight in LB, sub-cultured (1:50 dilution) in LB or serum for 2.5 h at 37 °C, and then washed and resuspended in PBS to a final $OD600_{nm}$ of 0.4. Approximately 2×10^7 cells were returned to 2 mL fresh LB or serum containing 75 μM DMACA-Olfox in the presence or absence of 20 μg mL^{-1} *A. baumannii* efflux-pump inhibitor AE0003 and incubated for 30 min at 37 °C. Cells were then washed with PBS and fixed with 4% formaldehyde in PBS for 30 min. After fixing, cells were washed in PBS, added to an 8 well-chambered coverglass (Nunc Lab-Tek, Nunc, Rochester, NY, USA), and stored at 4 °C for microscopy using a Nikon A1R confocal microscope, excited with a 405_{nm} laser, and resulting images were analyzed using Imaris 8 software (Oxford Instruments, Abingdon, UK).

3.12. RNA Sequencing

RNA sequencing was performed to compare the transcriptomes of *A. baumannii* strains 98-37-09 and 98-37-09 YhaK grown in human serum supplemented with 0.15 μg mL^{-1} levofloxacin. To do so, in duplicate, each strain was grown to the mid-exponential phase in human serum, and total bacterial RNA was isolated from each sample using Qiagen RNeasy kits. Library preparation and Next Generation Sequencing were performed at CD Genomics

(Shirley, New York, NY, USA). In brief, RNA concentration and quality were measured using Agilent 2100 Bioanalyzer and normalized, and rRNA was depleted using Ribo-Zero rRNA removal kits and fragmented. To construct sequencing libraries, cDNA synthesis was performed, the 3′ ends were adenylated, and adaptors were ligated. Paired-end sequencing was performed with Illumina NovaSeq PE150 (~20 million reads/sample, Illumina, San Diego, CA, USA). Processed/cleaned reads were then mapped to the *A. baumannii* ATCC 17978 reference genome (ASM1542v1) using Hierarchial Indexing for Spliced Alignment of Transcripts software (HISAT2 version 2.2.1) [34]. Transcripts were quantified, and gene expression levels were determined using Cufflinks software (v 2.2.1). Differentially Expressed Genes (DEGs) were classified based on their Gene Ontology (GO) functional terms and KEGG pathway enrichment analysis using the topGO software (Version 2.8). Raw sequences were deposited at the National Center for Biotechnology Information Sequence Read Archive (NCBI-SRA), under gene expression omnibus accession number GSE190119.

3.13. Statistics

Statistical analysis of flow cytometry and confocal microscopy data were performed using Graphpad prism 9, one-way Analysis of Variance (ANOVA), $p \leq 0.05$, whereas ethidium bromide and cell membrane permeability data were analyzed using Graphpad prim 9, Student's *t*-test $p \leq 0.05$. Differential Gene Expression Analysis was conducted using DESeq software (v 3.14). A false detection rate (FDR) ≤ 0.05 and the absolute value of Log_2 ratio ≥ 1 were used as the default threshold to identify significant DEGs.

4. Results and Discussion

Identification of *A. baumannii* mutants with impaired serum-associated antibiotic resistance.

Clinically defined antibiotic-susceptible *A. baumannii* strains have the remarkable ability to induce an expansive repertoire of drug efflux pumps during growth under physiological salt conditions, e.g., in pulmonary surfactant or human serum, which mediate resistance to physiologically achievable antibiotic concentrations within a patient [3–7]. More directly, and as previously shown, measures of cellular survival of the antibiotic susceptible strain 98-37-09 after incubation in conventional Luria-Bertani (LB) supplemented with various concentrations of minocycline, a front-line *A. baumannii* treatment option, revealed the strain to be susceptible to >1 μg mL^{-1} minocycline (Figure 1A) [35–43]. Conversely, the strain exhibited resistance to 2–4 μg mL^{-1} of the antibiotic during growth in human serum, correlating to or exceeding patient minocycline peak serum levels (~2.18 ± 0.44 μg mL^{-1} after 100 mg b.i.d. for 7–10 days) [44]. Moreover, the resistance phenotype was lost during growth in serum supplemented with the efflux pump inhibitor phenylalanyl arginyl β-naphthylamide (PaβN), indicating that serum-associated minocycline resistance is efflux-dependent (Figure 1A). This phenomenon, termed adaptive efflux mediated resistance (AEMR), has also been previously shown to modulate *A. baumannii* resistance to fluoroquinolones, aminoglycosides, and the glycylcline antibiotic, tigecycline, and is thought to account for one mechanism by which laboratory-defined antibiotic-susceptible strains resist antibiotic treatment within a patient [24].

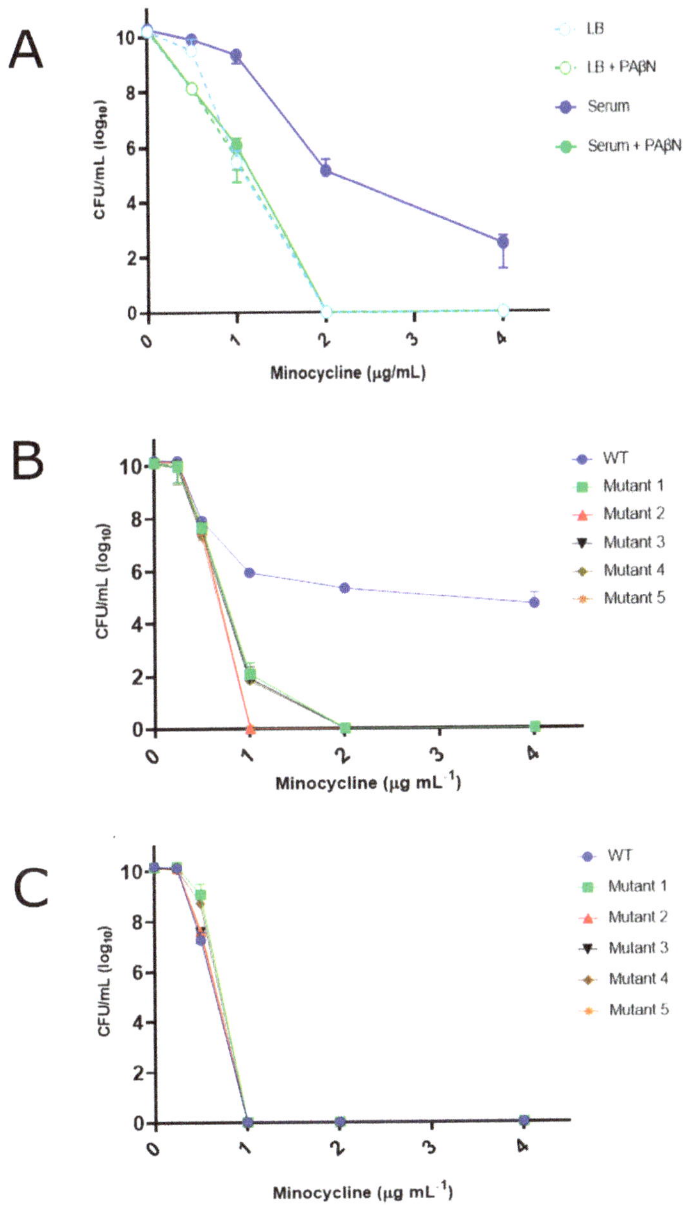

Figure 1. Growth phenotype of *A. baumannii* strains in Luria Bertani and human serum supplemented with minocycline. Plotted are the colony forming units (CFUs) of *A. baumannii* strain 98-37-09 (WT) during growth in LB media or human serum supplemented with 0, 0.25, 0.5, 1, 2, or 4 μg mL^{-1} minocycline +/− the efflux pump inhibitor PaβN (Panel **A**). Putative AEMR mutants following incubation in human serum (Panel **B**) or LB media (Panel **C**) supplemented with 0, 0.25, 0.5, 1, 2, or 4 μg mL^{-1} minocycline; standard deviations shown.

To identify genetic determinants that modulate AEMR, a strain 98-37-09 transposon mutant library was screened for members that exhibited a loss of minocycline AEMR phenotype during growth in serum. To do so, individual members of the library were inoculated in human serum supplemented with 1 μg mL^{-1} minocycline (1× MIC in LB; 0.5× MIC in serum), and growth was monitored. We predicted that library members with proficient serum-inducible AEMR response would grow in the presence of the antibiotic, whereas mutants with a deficient AEMR response would not. Among the 6000 mutants evaluated, 5995 library members exhibited minocycline resistance during serum growth, suggesting that their AEMR process(es) were intact. Conversely, 5 transposon library members (0.1%) exhibited a loss of growth phenotype when cultured serum in the presence of minocycline, but normal growth in serum lacking minocycline. Indeed, as shown in Figure 1B, dose escalation studies measuring the viability of strain 98-37-09 and each of the five mutants in human serum supplemented with 0, 0.25, 0.5, 1.0, 2.0, or 4.0 μg mL^{-1} minocycline revealed that while wild type 98-37-09 cells resisted the antimicrobial properties of the antibiotic, each of the mutants exhibited a loss of resistance phenotype at minocycline concentrations ≥ 1–2 μg mL^{-1}, indicating their AEMR response was impaired. Further, dose escalation testing in conventional LB media (Figure 1C) revealed that the susceptibility profile of each mutant of interest was virtually identical to that of wild-type 98-37-09 cells suggesting that the apparent loss of AEMR phenotype of each mutant in serum is not likely to be merely due to mutation of a gene that non-specifically leads to minocycline hypersusceptibility. Taken together, we considered that a subset of these mutants may harbor a transposon insertion within a genetic determinant that contributes to the *A. baumannii* AEMR response. Inverse PCR determined the transposon insertion site for each of the five mutants to be located within: a putative RND efflux pump protein (A1S_0008), a putative YhaK pirin-like protein (A1S_3277), isopropylmalate dehydratase (isomerase) subunit LeuD (A1S_0417), a putative ATP binding protein (A1S_0344), and a hypothetical protein (A1S_3826). These genes include three metabolic activity-associated genes, one efflux pump, and one pirin-like protein. The YhaK pirin-like protein possessed the strongest potential for a greater understanding of AEMR and thus was studied in more detail.

4.1. YhaK Impacts A. baumannii Serum-Inducible Efflux Properties

We were particularly intrigued that mutation of a putative YhaK pirin-like protein (A1S_3277) may be associated with an impaired AEMR response and chose to focus attention on this locus because pirin-like proteins have been associated with regulatory functions [45,46]. Thus, we considered that YhaK may play a key regulatory role in *A. baumannii* AEMR. To further evaluate whether YhaK contributes to AEMR, complementation was used to assess whether mutation of the pirin-like protein, as opposed to an unappreciated mutation within the strain, was responsible for the mutant's apparent loss of AEMR phenotype. Accordingly, the *A. baumannii* YhaK putative open reading frame and corresponding predicted promoter region from strain 98-37-09 were cloned into plasmid pWH1266 to create pYhaK and transformed into YhaK-mutant cells. Because the plasmid backbone contains a tetracycline resistance determinant that would complicate minocycline resistance analyses, we switched to evaluating the AEMR phenotype of the wild-type, YhaK-mutant, and complementation strain using another antibiotic, levofloxacin, which has been previously been shown to be affected by AEMR [26].

Consistent with previous results, testing revealed that wild-type 98-37-09 cells were susceptible to ≥0.125 μg mL^{-1} levofloxacin during growth in LB media, but resistant to a ≥4-fold increase in the antibiotic (0.5 μg mL^{-1}) during AEMR-inducing growth in human serum (Figure 2A) [26]. Conversely, the YhaK-mutant exhibited a loss of resistance phenotype, displaying susceptibility to 0.5 μg mL^{-1} levofloxacin during serum growth. The mutant's apparent loss of levofloxacin resistance during growth in human serum was unaffected by the presence of the shuttle vector pWH1266 but could be rescued by plasmid pYhaK harboring a wild-type copy of the gene, suggesting that the YhaK pirin-like protein contributes to *A. baumannii* AEMR (Figure 2A). Attempts to create a clean strain 98-37-09 *yhaK* deletion were unsuccessful. Thus, to verify the effects of YhaK on

AEMR in an independent strain background, we obtained *A. baumannii* strains AB5075-UW (wild type) and an isogenic YhaK-mutant, AB5075-YhaK, from the Manoil transposon mutant library [30] and assessed their levofloxacin susceptibility profiles during AEMR-inducing serum growth conditions. Results revealed that while the wild-type strain, AB5075-UW was intrinsically more resistant to levofloxacin than strain 98-37-09, and it was susceptible to ≥16 μg mL^{-1} levofloxacin during growth in LB media and exhibited a ≥4-fold increase in resistance to (64 μg mL^{-1}) during growth in human serum suggesting it undergoes AEMR (Figure 2B). Conversely, the YhaK-mutant strain exhibited loss of resistance to 64 μg mL^{-1} levofloxacin during serum growth, which could be restored by complementation with pYhaK harboring a wild-type copy of the *yhaK* gene but not by the plasmid vector alone. Taken together, these results indicated that the YhaK pirin-like protein is required for levofloxacin resistance during serum growth in at least two strain backgrounds, presumably by participating in the AEMR response.

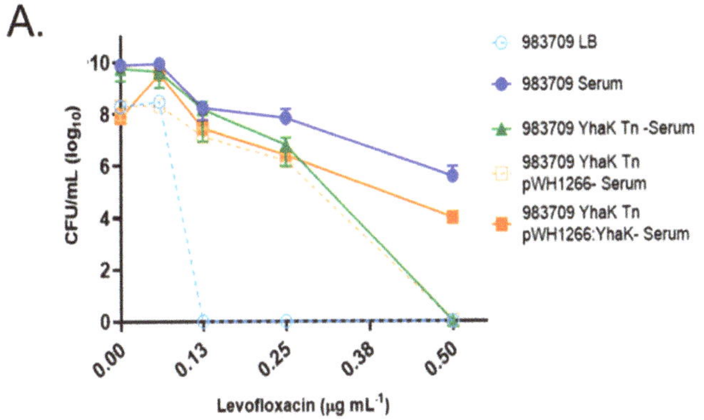

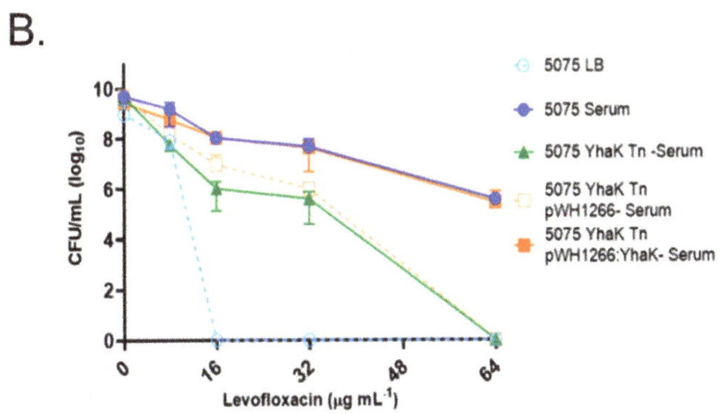

Figure 2. Growth phenotype of *A. baumannii* strains in Luria Bertani and human serum supplemented with levofloxacin. (**A**) Plotted are the colony-forming units (CFUs) of *A. baumannii* strain 98-37-09 (WT), YhaK transposon mutant (YhaK Tn), and YhaK transposon mutant containing plasmid pWH1266 or pWH1266::YhaK following incubation in human serum supplemented with 0, 0.06, 0.125, 0.25, or 0.5 μg mL^{-1} levofloxacin; standard deviations shown. (**B**) Plotted are the colony-forming units (CFUs) of *A. baumannii* strain AB5075-UW (WT), YhaK transposon mutant (YhaK Tn), and YhaK transposon mutant containing plasmid pWH1266 or pWH1266::YhaK following incubation in human serum supplemented with 0, 8, 16, 32, or 64 μg mL^{-1} levofloxacin; standard deviations shown.

To determine whether the apparent loss of AEMR phenotype of YhaK-mutant cells does indeed correspond with a loss of the strain's efflux properties, standard ethidium bromide (EtBr) efflux assays were used to compare the efflux capacity of wild-type 98-37-09 and YhaK-mutant cells during growth in LB (non-AEMR) and human serum (AEMR inducing) conditions. The premise for such EtBr assays is that efflux-deficient cells will accumulate ethidium bromide and display high corresponding fluorescence in comparison to efflux-proficient cells that extrude the dye [47]. As previously reported, EtBr assays of wild-type cells grown in LB and serum (Figure 3A), revealed the strain exhibited increased EtBr accumulation during growth in LB media (3332.14 $\pm$ 38.79 Relative Fluorescent units; RFUs) in comparison to growth in human serum (2576.17 $\pm$ 40.88 RFU), confirming earlier studies that showed serum-induced AEMR antibiotic resistance correlates with increased efflux capacity [26]. Conversely, studies of the YhaK-mutant indicated that it exhibits higher ethidium bromide accumulation (2959.5 $\pm$ 51.47 RFU) during serum growth in comparison to wild-type cells, suggesting that the apparent loss of AEMR due to *yhaK*-mutation correlates with a loss of serum-inducible efflux capacity (Figure 3A). To distinguish whether the observed differences in EtBr accumulation between serum-grown wild-type and mutant cells may be attributable to alterations in cell membrane permeability, as opposed to differences in the efflux properties of the two strains, standard 1-*N*-phenylnapthylamine (NPN) cell permeability assays were performed on serum-grown wild-type and mutant cells. As shown in Figure 3B, no significant difference in NPN cell permeability was observed between the two strains during serum growth; the detergent triton X-100 was used as a positive control to ensure the assay was indeed capable of measuring increased cell permeability of both strains in the presence of human serum. Taken together, these results indicate that the apparent loss of AEMR associated with the mutation of *yhaK* is likely to be due to limited efflux potential as opposed to compromised membrane permeability.

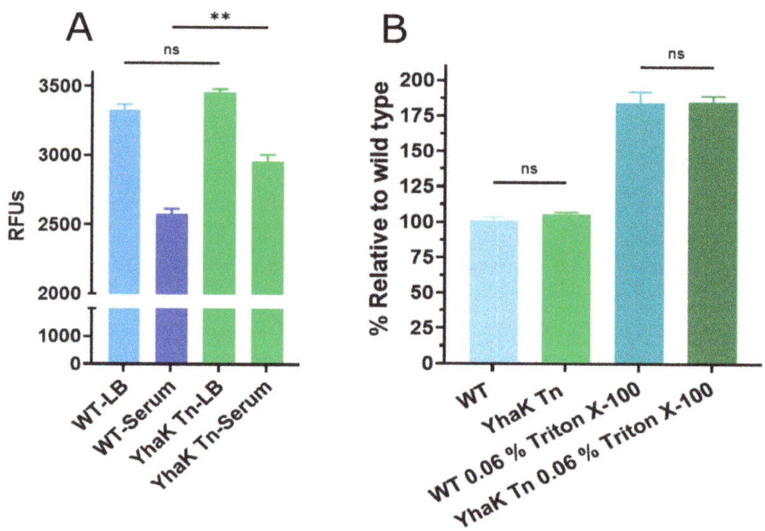

Figure 3. Ethidium bromide accumulation and 1-*N*-phenylnapthylamine cell permeability assays. (Panel **A**) Plotted are the relative fluorescent units (RFUs) of ethidium bromide-treated *A. baumannii* strains 98-37-09 (WT) and isogenic YhaK-mutant (YhaK Tn) following growth in LB or human serum. (Panel **B**). Plotted are the RFUs of WT and YhaK Tn cells following growth in human serum and treatment with the cell permeability dye 1-*N*-phenylnapthylamine in the absence or presence of 0.06% Triton X-100 detergent; standard deviations and significance shown (Student's *t*-test, ** $p \leq 0.01$; not significant (ns)).

4.2. YhaK Impacts *A. baumannii* Serum-Inducible Adaptive Efflux-Mediated Antibiotic Resistance

While ethidium bromide has historically served as an excellent surrogate for measuring the overall efflux properties of bacterial cells, we set out to more conclusively assess whether YhaK mutation affects *A. baumannii* antibiotic extrusion. Accordingly, a fluorescent isomer of levofloxacin, ofloxacin, was created (DMACA-Oflox; Supplemental Figure S1A) as a tool to more directly measure AEMR-mediated antibiotic accumulation within wild-type and YhaK-mutant cells. We chose DMACA because it is a low molecular weight fluorescent molecule that has successfully been shown to measure bacterial quinolone internalization and efflux, while also retaining antibiotic activity [48]. As a prerequisite for doing so, we assessed whether wild-type cells exhibited AEMR-associated resistance to ofloxacin and/or the fluorescent derivative.

As shown in Figure 4A, wild-type 98-37-09 cells were susceptible to $\geq$0.25 µg mL^{-1} ofloxacin during growth in LB but resistant to $\geq$1 µg mL^{-1} of the antibiotic during growth in serum, suggesting that ofloxacin is impacted by AEMR. Conversely, such resistance was ablated for the mutant during growth in human serum. More directly, the YhaK-mutant strain was found to be susceptible to $\geq$0.25 µg mL^{-1} ofloxacin during growth in LB but showed a loss of resistance to 1 µg mL^{-1} of the antibiotic during serum growth, indicating that YhaK partially impacts serum growth associated ofloxacin resistance (Figure 4A). Similarly, while fluorescence-labeling seemed to reduce ofloxacin's antimicrobial potency, wild-type cells displayed resistance to $\geq$200 µg mL^{-1} DMACA-Oflox during serum growth whereas serum-grown YhaK-mutant cells exhibited a loss of resistance to 200 µg mL^{-1} DMACA-Oflox (Figure 4B). Moreover, the YhaK-mutant phenocopied DMACA-Oflox-treated serum-grown wild-type cells in the presence of the known *A. baumannii* AEMR inhibitor, AE0003 (structure provided in Supplemental Figure S1B), indicating that mutation of the strain's YhaK protein renders them AEMR-deficient (Figure 4C) [33].

4.3. YhaK-Mutant Cells Accumulate DMACA-Oflox

Confocal microscopy was used to determine whether the apparent loss of DMACA-Oflox adaptive efflux-mediated resistance in YhaK-mutant cells did indeed correlate with increased cellular accumulation of the ofloxacin derivative in comparison to AEMR-proficient wild-type cells. As expected, confocal microscopy revealed that wild-type cells accumulate DMACA-Oflox during growth in LB media (448 $\pm$ 50 relative fluorescent units; RFU) but display a 15.4-fold reduction in accumulation (29 $\pm$ 4 RFU) during AEMR-inducing serum growth conditions (Figure 5A,B). Tests of the YhaK mutant revealed (Figure 5C,D) that the strain accumulates the antibiotic-derivative to wild-type levels during growth in LB (484 $\pm$ 65 RFU) yet only results in a 5.7-fold reduction in accumulation (84 $\pm$ 16 RFU) during growth in human serum, suggesting that the mutant's loss of DMACA-Oflox AEMR correlates with increased intracellular antibiotic accumulation. Figure 5E quantifies the average confocal microscopy fluorescence signals for all assay conditions. Moreover, as shown in Figure 6, the reduced DMACA-Oflox accumulation of serum-grown wild-type cells could, at least in part, could be reversed by the *A. baumannii* efflux pump inhibitor AE0003, indicating that DMACA-Oflox allows measures of the organism's adaptive efflux-mediated resistance.

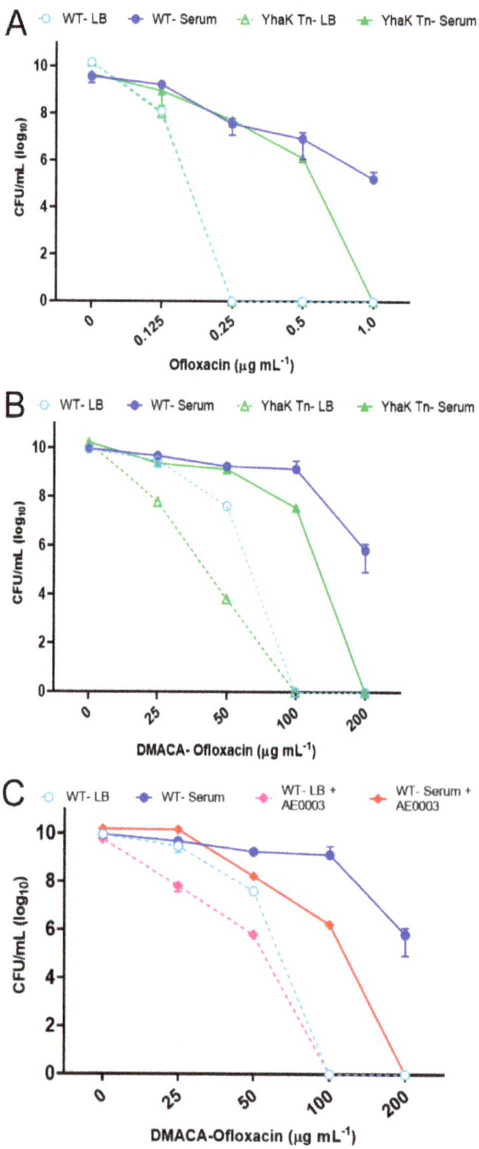

Figure 4. Growth phenotype of *A. baumannii* strains in Luria Bertani and human serum supplemented with ofloxacin or DMACA-ofloxacin. Plotted are the colony-forming units (CFUs) of *A. baumannii* strain 98-37-09 (WT) and YhaK Tn mutant (Yhak Tn) following incubation in LB media or human serum supplemented with the indicated concentration of the antibiotic olfoxacin (Panel **A**) or DMAC-olfoxacin (Panel **B**). (Panel **C**) Plotted are CFUs of WT cells following incubation in LB or human serum treated with DMACA-ofloxacin in the absence or presence of the efflux pump inhibitor AE0003; standard deviations are shown.

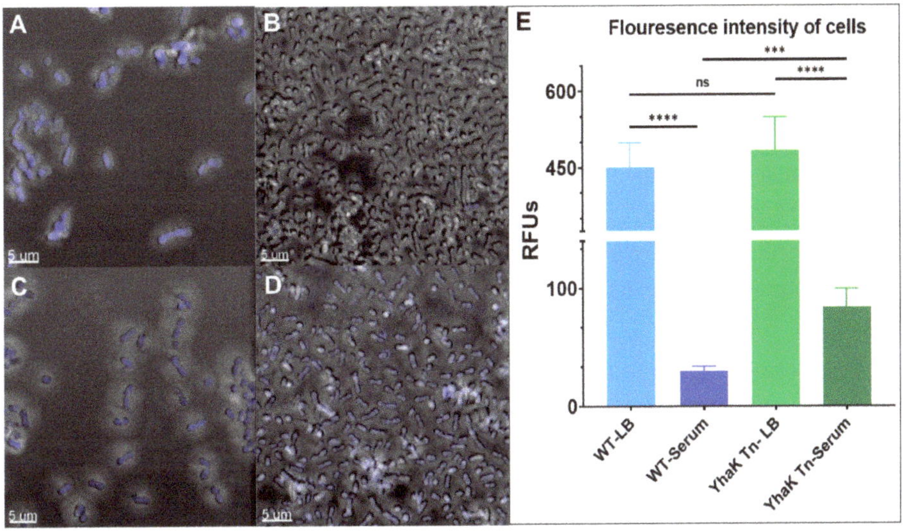

Figure 5. Confocal microscopy of DMACA-Ofloxacin treated wild-type and YhaK Tn-mutant cells following growth in LB or human serum. Shown are stacked confocal images of DMACA-Oflox-treated WT cells following growth in LB (Panel **A**) or human serum (Panel **B**) or YhaK-mutant cells following growth in LB (Panel **C**) or human serum (Panel **D**). (Panel **E**). Averaged fluorescence intensity (n = 20) graphed; standard deviation shown; statistical difference (One-way ANOVA, *** $p \leq 0.001$, **** $p \leq 0.0001$; not significant (ns)).

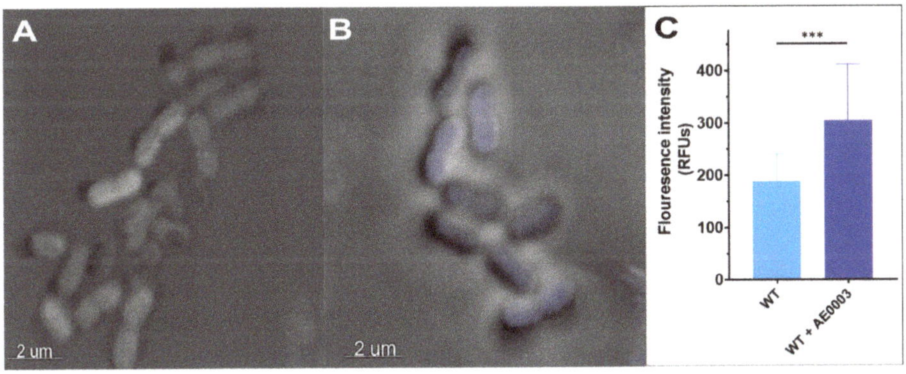

Figure 6. Confocal microscopy of DMACA-Ofloxacin treated wild-type cells in the presence of the efflux pump inhibitor AE0003. Stacked confocal images of DMACA-Oflox treated serum grown WT cells in the absence (Panel **A**) or presence (Panel **B**) of the efflux pump inhibitor AE0003. (Panel **C**). Averaged fluorescent signal of untreated and AE0003-treated serum-grown WT cells; standard deviation shown (n = 20); *** indicates statistical difference (Student's t-test, $p \leq 0.001$).

Similarly, flow cytometry studies indicated that YhaK-mutant cells exhibited increased DMACA-Oflox accumulation during growth in human serum in comparison to wild-type cells. More directly, wild-type and YhaK-mutant cells grown in LB-supplemented 75 μM DMACA-Oflox produced side scatter distributions of $\geq 10^4$ RFU indicative of high accumulation of the antimicrobial, plotted as light blue and light green within Figure 7A,B, respectively. However, during AEMR-inducing serum growth, wild-type cells exhibited a significant reduction in DMACA-Oflox association, resulting in a cell population shift with approximately a 10-fold decrease in RFU (comparison of light blue to dark blue) resulting in a phenotype that could partially be rescued by cellular treatment with the AEMR inhibitor AE0003 (Figure 7C,D), indicating flow cytometry allows measures of the organism's AEMR phenotype. In direct agreement with confocal microscopy, tests of the serum-grown YhaK-mutant cells revealed an intermediate cell population shift of approximately 5-fold reduction in RFU units (comparison of light green to dark green), suggesting that the mutant has increased DMACA-Oflox association in comparison to wild-type cells during growth in human serum (Figure 7A,B).

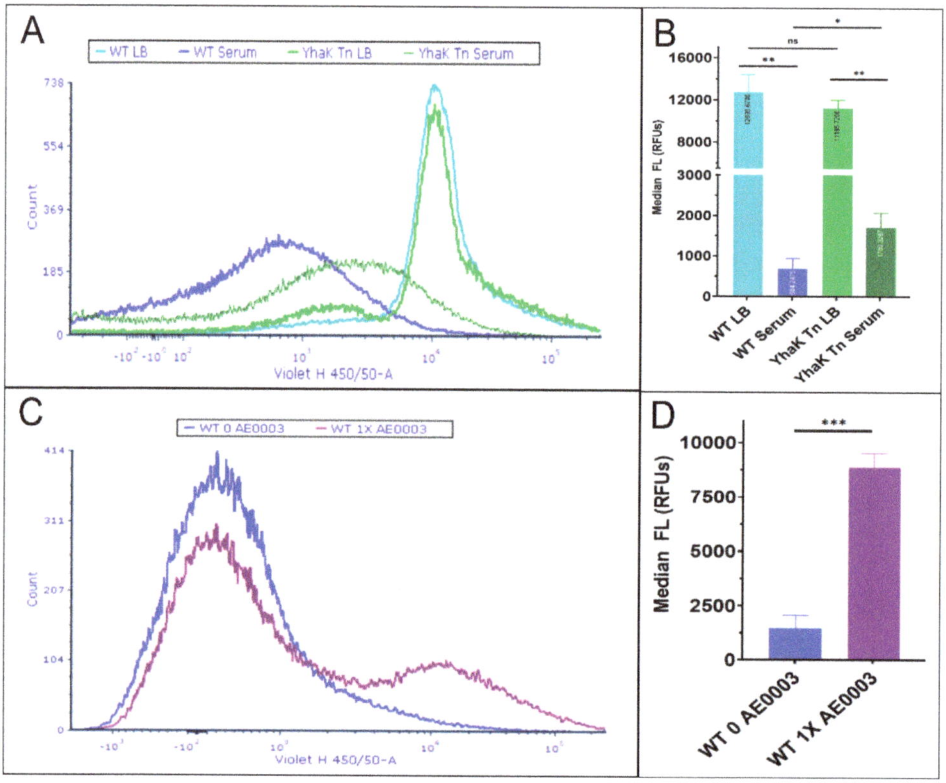

Figure 7. Flow cytometry of DMACA-Ofloxacin-treated cells. (Panel **A**) Plotted are the intensity measures of DMACA-Ofloxacin-treated WT cells during growth in LB (light blue) or human serum (dark blue) or YhaK-mutant (Yhak-Tn) cells following LB (light green) or serum (dark green) growth, with averaged median fluorescence from duplicate studies shown in (Panel **B**). (Panel **C**) Plotted are the fluorescence intensity and side scatter of DMACA-Oflox treated serum grown WT cells in the absence (dark blue) or presence (purple) of the *A. baumannii* efflux pump inhibitor AE0003, with averaged median fluorescence from duplicate studies shown in (Panel **D**). Standard deviation shown; statistical significance (One-way ANOVA, * $p \leq 0.05$, ** $p \leq 0.01$, *** $p \leq 0.001$ and not significant (ns)).

4.4. RNA Sequencing of Wild-Type and YhaK-Mutant Cells during Growth in Human Serum

Recognizing that pirin-like proteins display regulatory effects and hence YhaK's putative contributions to AEMR may be regulatory in nature, we set out to assess the protein's cellular effects using RNA sequencing. As a first step toward doing so, we assessed the protein's expression within wild-type cells during growth in LB and human serum, both in the absence and presence of levofloxacin using quantitative RT-PCR. Results (Supplemental Figure S2) revealed that the YhaK gene is upregulated approximately

2-fold in cells grown in human serum in comparison to growth in LB media. Further, the presence of the antibiotic stimulated gene transcription approximately 4-fold, suggesting that its regulatory effects may be optimally measured by comparing the transcriptomes of wild-type and YhaK-mutant cells during growth in human serum supplemented with levofloxacin.

Comparison of the gene expression properties of wild-type and YhaK-mutant cells during these conditions revealed that 333 genes were upregulated and 320 genes were down-regulated at least 2-fold (FDR $\leq$ 0.05) in wild-type cells in comparison to the mutant, suggesting that mutation of the YhaK protein results in pleiotropic effects. To focus attention on defining the potential mechanism(s) by which YhaK may modulate adaptive efflux-mediated antibiotic resistance, Figure 8 provides an overview of the predominant KEGG gene ontology groups differentially expressed in a YhaK-dependent manner. Within YhaK expressing wild-type cells, genes involved in transcription, cell signaling, and post-translational modification were highly expressed in comparison to mutant cells, as were systems belonging to numerous transport modalities including nucleotides, inorganic ions, co-enzymes, and amino acids. Among the most repressed within wild-type cells where genes are associated with translation machinery, secondary metabolites, lipid transport, and energy production.

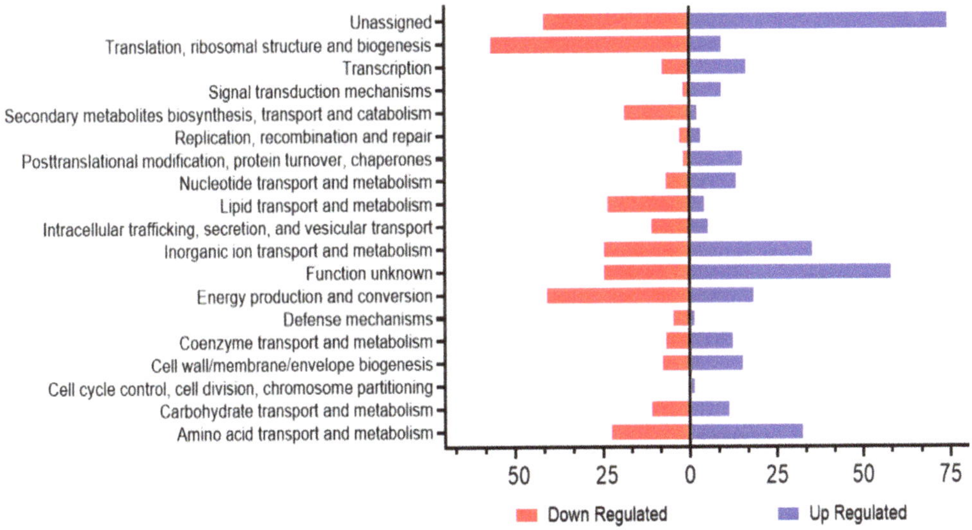

Figure 8. *A. baumannii* WT and YhaK Tn Transcriptome analysis: KEGG gene ontology groups. A number of genes displaying > 2-fold alteration in wild-type compared to YhaK-mutant cells by gene ontology.

Further assessment of individual components of these ontology groups provided several plausible mechanisms by which YhaK may modulate adaptive efflux-mediated resistance. In perhaps the most straightforward scenario, one could imagine that YhaK stimulates the expression of antibiotic drug efflux pumps during growth in human serum, thereby directly accounting for an AEMR phenotype. While the majority of previously characterized *A. baumannii* antibiotic efflux pumps were not differentially expressed between the two strains, 15 transport pumps, 11 of which are predicted, have been associated with antibiotic resistance within other organisms, where they were found to be upregulated at least 2-fold in wild-type cells during serum growth in the presence of levofloxacin in comparison to YhaK-mutant cells (Figure 9). These include a putative MATE drug efflux pump (3.5×; locus A1S_3420), four MFS pump proteins (2.1–8.2×; A1S_0801, A1S_0802, A1S_2326 and A1S_2327), four ATP-binding cassette pump proteins (3–4.2×; loci A1S_1611, A1S_1613, A1S_1786, and A1S_2299), and ammonium transporters (4.8–5.2×; loci A1S_021 and A1S_2748; ref. [49]). In a second possibility, YhaK may be responsible for activating cellular efflux capability post-translationally. Indeed, studies have revealed that drug efflux pumps can be post-translationally regulated and/or require modification to become activated, and as shown in Figure 8, post-translational regulatory processes are among the most highly upregulated within wild-type as opposed to YhaK-mutant cells during serum growth in the presence of levofloxacin [50]. Third, efflux pumps require energy for activity, and two ammonium transporters (above) were found to be expressed in a YhaK-dependent manner. A fourth possibility is that YhaK's affects serum-associated antibiotic resistance in a mechanism that does not directly involve modulating cellular efflux. To that end, results revealed that ClpP protease is upregulated 2.3-fold within wild-type, as opposed to YhaK-mutant cells. ClpPX has been shown to affect antibiotic tolerance during stress conditions in other Gram-negative pathogens, such as *E. coli* [51].

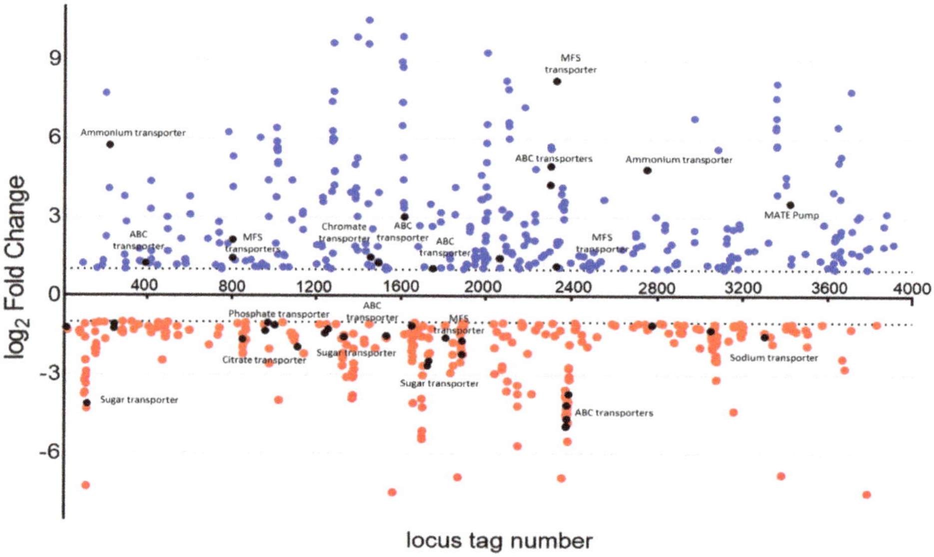

Figure 9. *A. baumannii* WT and YhaK Tn Transcriptome analysis: differentially expressed genes. Scatter plot of differentially expressed genes within wild-type and YhaK-mutant cells during grown in serum supplemented with levofloxacin; $\log_2$ fold change in expression for each gene was plotted against gene locus tag number. Loci determined to be upregulated (greater transcript titers) in wild-type cells in comparison to YhaK-mutant cells are labeled in blue, whereas downregulated genes are labeled red. Loci annotated to code for known or putative transporters are labeled in black; known transporters are also labeled.

Regardless of the molecular mechanism(s), taken together, these results suggest that YhaK is likely to contribute to *A. baumannii*'s AEMR phenotype. *A. baumannii* YhaK (accession # COG1741) is predicted to code for a 290-amino acid gene product that is annotated as pirin-like protein within the cupin superfamily based on primary sequence alignment. BLAST search indicates that the protein is highly conserved across 176 publicly available *A. baumannii* genomes ($\geq$97% amino acid identity) and shares 55–70% amino acid identity with proteins of other Gram-negative pathogens including *P. aeruginosa*, *K. pneumonia*, *Yersinia pestis*, and *Salmonella enterica*, suggesting that its cellular functions may be conserved across species. Until recently, pirin-like protein function(s) have been poorly understood, with reports spanning both enzymatic and non-enzymatic activities. Arguably, most information about the protein family derives from studies of human pirin (hPirin) and the *E. coli* pirin protein, YhhW. hPirin has been shown to bind transcription factors NF-I and Bcl-3 in vivo, suggesting that pirins act as transcription co-factors to regulate gene transcription [52,53]. In the case of the latter, hPirin is thought to stabilize quaternary complexes between Bcl-3, NF-κB, and promoter sequences [54]. To our knowledge, no studies have evaluated whether *E. coli* YhhW displays regulatory effects, yet the tertiary structures of both human pirin and YhhW have been determined and are similar to quercetin 2,3-dioxygenase, which led to the finding that both hPirin and YhhW display quercetinase activity [55]. It is not clear whether the hPirin's enzymatic activity is associated, either directly or indirectly, with the protein's regulatory function(s). More recently, it was found that although the *E. coli* YhaK pirin-like protein shares overall topology with hPirin and YhhW, it does not seem to exhibit quercetinase activity, presumably due to the absence of three conserved pirin-like protein amino acid residues (His57, His59, and His101) that are thought to govern enzymatic function [56]. Hence, it has been proposed that the *E. coli* protein should be reannotated solely as a bicupin to distinguish it from the *E. coli* YhhW pirin-like protein [56].

Alignments of the *A. baumannii* YhaK protein that is the focus of these studies indicates that the protein best fits and shares the conserved residues of hPirin and YhhW pirin family hypothesized to mediate quercetinase activity, suggesting that the protein may display both regulatory and enzymatic activities that have been associated with other pirins (Supplemental Figure S3). To that end, RNA sequencing of wild-type and YhaK-mutant cells during AEMR-inducing growth conditions revealed that *A. baumannii* YhaK is required for wild-type levels of expression of multiple transporter systems including those that are ostensibly associated with antibiotic efflux, such as a multidrug and toxic compound extrusion (MATE) family protein and several ATP-binding cassette (ABC) pumps. It remains to be seen whether the protein's regulatory effects are direct or indirect. Interestingly, YhaK did not seem to affect the expression of the RND-pump protein identified in this study (A1S_0008) or pumps that have been previously characterized to mediate antibiotic efflux. It will be interesting to learn whether point mutations within YhaK residues that are predicted to be required for quercetinase activity affect the protein's regulatory effects.

While it is intriguing to consider that *A. baumannii* YhaK's contribution to AEMR is via transcriptional regulation of efflux pumps, other regulatory effects were also noted, as described above, suggesting that mutation of the protein leads to pleiotropic effects, and hence YhaK's role in modulating AEMR may be indirect. Regardless of the mechanism, the results presented indicate that the protein is a determinant of the organism's serum-inducible AEMR response to at least two classes of antibiotics, the fluoroquinolones and tetracyclines. As such, one can imagine that small molecule inhibitors of YhaK activity may have therapeutic utility as adjuvants, which improve the therapeutic potential of these antibiotics toward *A. baumannii*. In that regard, it will be important to evaluate whether YhaK is required for AEMR toward other classes of antibiotics and whether these effects are conserved across contemporary *A. baumannii* clinical isolates in future studies.

The transposon mutant library screen used to identify YhaK also revealed that mutation of four other loci, including a putative RND efflux pump protein, isopropylmalate dehydratase, a putative ATP binding protein, and a hypothetical protein, may also affect *A. baumannii*'s AEMR response. Ostensibly, the RND efflux pump system may play a direct role by extruding antibiotics during AEMR-inducing conditions. While the putative contributions of the other gene products are more obscure, we expect that as components of the system are identified, verified, and characterized, their roles will become clearer. Admittedly, a limitation of the transposon screening approach used is that it, by definition, excludes evaluating the contributions of essential gene products. In that regard, YhaK may serve as a biochemical and/or genetic tool to identify additional components of *A. baumannii* AEMR.

Supplementary Materials: The following supporting information can be downloaded at: https://www.mdpi.com/article/10.3390/antibiotics12071173/s1, Figure S1: Chemical structures; Figure S2: Quantitative real time PCR measures of YhaK transcription; Figure S3: ClustalW alignment of *Escherichia coli* pirin-like protein YhaK, *E. coli* 2,3 dioxygenase YhhW, human pirin protein (hPirin), and *A. baumannii* YhaK pirin-like protein (A1S_3277).

Author Contributions: Conceptualization, D.F. and P.M.D.; Methodology, M.Y. and M.C.; Formal analysis, W.L.J.; Investigation, M.Y., M.C., C.B. and X.C.; Writing—original draft, M.Y., M.C., D.F. and P.M.D.; Writing—review & editing, D.F. and P.M.D.; Funding acquisition, D.F. and P.M.D. All authors have read and agreed to the published version of the manuscript.

Funding: This research was funded by National Institute of Allergy and Infectious Diseases (Grant No. AI175024).

Institutional Review Board Statement: Not applicable.

Informed Consent Statement: Not applicable.

Acknowledgments: M.Y. and P.M.D. are partially supported by the University of Rochester Technology Development Fund (TDF). This work was also partially supported by the National Institute of Allergy and Infectious Disease (NIAID) award AI175024 (P.M.D. and D.F.). *A. baumannii* 5075 and (ABUW_0207-119::T26) were obtained from Manoil Lab at the University of Washington (NIH P30 DK089507).

Conflicts of Interest: P.M.D. is founder of Arcum Therapeutics and Arcum Vision, two small molecule antimicrobial biotechnology companies. The work associated with these studies is independent of small molecule research.

References

1. Verma, P.; Tiwari, M.; Tiwari, V. Efflux pumps in multidrug-resistant *Acinetobacter baumannii*: Current status and challenges in the discovery of efflux pumps inhibitors. *Microb. Pathog.* **2021**, *152*, 104766. [CrossRef] [PubMed]
2. Fernández, L.; Breidenstein, E.B.; Hancock, R.E. Creeping baselines and adaptive resistance to antibiotics. *Drug Resist. Updat.* **2011**, *14*, 1–21. [CrossRef] [PubMed]
3. Hornsey, M.; Ellington, M.J.; Doumith, M.; Thomas, C.P.; Gordon, N.C.; Wareham, D.W.; Quinn, J.; Lolans, K.; Livermore, D.M.; Woodford, N. AdeABC-mediated efflux and tigecycline MICs for epidemic clones of *Acinetobacter baumannii*. *J. Antimicrob. Chemother.* **2010**, *65*, 1589–1593. [CrossRef]
4. Metan, G.; Alp, E.; Yildiz, O.; Percin, D.; Aygen, B.; Sumerkan, B. Clinical experience with tigecycline in the treatment of carbapenem-resistant Acinetobacter infections. *J. Chemother.* **2010**, *22*, 110–114. [CrossRef] [PubMed]
5. Peleg, A.Y.; Potoski, B.A.; Rea, R.; Adams, J.; Sethi, J.; Capitano, B.; Husain, S.; Kwak, E.J.; Bhat, S.V.; Paterson, D.L. *Acinetobacter baumannii* bloodstream infection while receiving tigecycline: A cautionary report. *J. Antimicrob. Chemother.* **2007**, *59*, 128–131. [CrossRef]
6. Rice, L.B. Challenges in identifying new antimicrobial agents effective for treating infections with *Acinetobacter baumannii* and *Pseudomonas aeruginosa*. *Clin. Infect. Dis.* **2006**, *43* (Suppl. S2), S100–S105. [CrossRef] [PubMed]
7. Hood, M.I.; Jacobs, A.C.; Sayood, K.; Dunman, P.M.; Skaar, E.P. *Acinetobacter baumannii* increases tolerance to antibiotics in response to monovalent cations. *Antimicrob. Agents Chemother.* **2010**, *54*, 1029–1041. [CrossRef]
8. Jacobs, A.C.; Sayood, K.; Olmsted, S.B.; Blanchard, C.E.; Hinrichs, S.; Russell, D.; Dunman, P.M. Characterization of the *Acinetobacter baumannii* growth phase-dependent and serum responsive transcriptomes. *FEMS Immunol. Med. Microbiol.* **2012**, *64*, 403–412. [CrossRef]
9. CDC. *Antibiotic Resistance Threats in the United States*; CDC: Atlanta, GA, USA, 2019; pp. 1–150.
10. O'Neill, J. *Tackling Drug-Resistant Infections Globally: Final Report and Recommendations*; Wellcome Trust and HM Government: London, UK, 2016.
11. Hidron, A.I.; Edwards, J.R.; Patel, J.; Horan, T.C.; Sievert, D.M.; Pollock, D.A.; Fridkin, S.K. NHSN annual update: Antimicrobial-resistant pathogens associated with healthcare-associated infections: Annual summary of data reported to the National Healthcare Safety Network at the Centers for Disease Control and Prevention, 2006–2007. *Infect. Control Hosp. Epidemiol.* **2008**, *29*, 996–1011. [CrossRef]
12. Gupta, V.; Ye, G.; Olesky, M.; Lawrence, K.; Murray, J.; Yu, K. Trends in resistant Enterobacteriaceae and Acinetobacter species in hospitalized patients in the United States: 2013–2017. *BMC Infect. Dis.* **2019**, *19*, 742. [CrossRef]
13. Abadi, A.T.B.; Rizvanov, A.A.; Haertle, T.; Blatt, N.L. World Health Organization Report: Current Crisis of Antibiotic Resistance. *BioNanoScience* **2019**, *9*, 778–788. [CrossRef]
14. Rice, L.B. Federal funding for the study of antimicrobial resistance in nosocomial pathogens: No ESKAPE. *J. Infect. Dis.* **2008**, *197*, 1079–1081. [CrossRef]
15. Fournier, P.E.; Vallenet, D.; Barbe, V.; Audic, S.; Ogata, H.; Poirel, L.; Richet, H.; Robert, C.; Mangenot, S.; Abergel, C.; et al. Comparative genomics of multidrug resistance in *Acinetobacter baumannii*. *PLoS Genet.* **2006**, *2*, e7. [CrossRef] [PubMed]
16. Yoon, E.J.; Balloy, V.; Fiette, L.; Chignard, M.; Courvalin, P.; Grillot-Courvalin, C. Contribution of the Ade Resistance-Nodulation-Cell Division-Type Efflux Pumps to Fitness and Pathogenesis of *Acinetobacter baumannii*. *mBio* **2016**, *7*, e00697-16. [CrossRef] [PubMed]
17. Nishino, K.; Yamaguchi, A. Analysis of a complete library of putative drug transporter genes in *Escherichia coli*. *J. Bacteriol.* **2001**, *183*, 5803–5812. [CrossRef]

18. Paulsen, I.T.; Sliwinski, M.K.; Saier, M.H., Jr. Microbial genome analyses: Global comparisons of transport capabilities based on phylogenies, bioenergetics and substrate specificities. *J. Mol. Biol.* **1998**, *277*, 573–592. [CrossRef]
19. Wan, Y.; Wang, M.; Chan, E.W.C.; Chen, S. Membrane Transporters of the Major Facilitator Superfamily Are Essential for Long-Term Maintenance of Phenotypic Tolerance to Multiple Antibiotics in *E. coli*. *Microbiol. Spectr.* **2021**, *9*, e0184621. [CrossRef]
20. Zhang, Y.; Xiao, M.; Horiyama, T.; Zhang, Y.; Li, X.; Nishino, K.; Yan, A. The Multidrug Efflux Pump MdtEF Protects against Nitrosative Damage during the Anaerobic Respiration in *Escherichia coli*. *J. Biol. Chem.* **2011**, *286*, 26576–26584. [CrossRef]
21. Adams, K.N.; Takaki, K.; Connolly, L.E.; Wiedenhoft, H.; Winglee, K.; Humbert, O.; Edelstein, P.H.; Cosma, C.L.; Ramakrishnan, L. Drug tolerance in replicating mycobacteria mediated by a macrophage-induced efflux mechanism. *Cell* **2011**, *145*, 39–53. [CrossRef]
22. Kristoffersen, S.M.; Ravnum, S.; Tourasse, N.J.; Økstad, O.A.; Kolstø, A.B.; Davies, W. Low concentrations of bile salts induce stress responses and reduce motility in Bacillus cereus ATCC 14579. *J. Bacteriol.* **2007**, *189*, 5302–5313. [CrossRef]
23. Urdaneta, V.; Casadesús, J. Adaptation of *Salmonella enterica* to bile: Essential role of AcrAB-mediated efflux. *Environ. Microbiol.* **2018**, *20*, 1405–1418. [CrossRef] [PubMed]
24. Fernández, L.; Hancock, R.E. Adaptive and mutational resistance: Role of porins and efflux pumps in drug resistance. *Clin. Microbiol. Rev.* **2012**, *25*, 661–681. [CrossRef] [PubMed]
25. Morrison, J.M.; Chojnacki, M.; Fadrowski, J.J.; Bauza, C.; Dunman, P.M.; Dudas, R.A.; Goldenberg, N.A.; Berman, D.M. Serum-Associated Antibiotic Tolerance in Pediatric Clinical Isolates of *Pseudomonas aeruginosa*. *J. Pediatric Infect. Dis. Soc.* **2020**, *9*, 671–679. [CrossRef]
26. Blanchard, C.; Barnett, P.; Perlmutter, J.; Dunman, P.M. Identification of *Acinetobacter baumannii* serum-associated antibiotic efflux pump inhibitors. *Antimicrob. Agents Chemother.* **2014**, *58*, 6360–6370. [CrossRef]
27. Jacobs, A.C.; Blanchard, C.E.; Catherman, S.C.; Dunman, P.M.; Murata, Y. An ribonuclease T2 family protein modulates *Acinetobacter baumannii* abiotic surface colonization. *PLoS ONE* **2014**, *9*, e85729. [CrossRef] [PubMed]
28. Jacobs, A.C.; Hood, I.; Boyd, K.L.; Olson, P.D.; Morrison, J.M.; Carson, S.; Sayood, K.; Iwen, P.C.; Skaar, E.P.; Dunman, P.M. Inactivation of phospholipase D diminishes *Acinetobacter baumannii* pathogenesis. *Infect. Immun.* **2010**, *78*, 1952–1962. [CrossRef]
29. Hunger, M.; Schmucker, R.; Kishan, V.; Hillen, W. Analysis and nucleotide sequence of an origin of DNA replication in *Acinetobacter calcoaceticus* and its use for *Escherichia coli* shuttle plasmids. *Gene* **1990**, *87*, 45–51. [CrossRef]
30. Gallagher, L.A.; Ramage, E.; Weiss, E.J.; Radey, M.; Hayden, H.S.; Held, K.G.; Huse, H.K.; Zurawski, D.V.; Brittnacher, M.J.; Manoil, C. Resources for Genetic and Genomic Analysis of Emerging Pathogen *Acinetobacter baumannii*. *J. Bacteriol.* **2015**, *197*, 2027–2035. [CrossRef]
31. Altschul, S.F.; Gish, W.; Miller, W.; Myers, E.W.; Lipman, D.J. Basic local alignment search tool. *J. Mol. Biol.* **1990**, *215*, 403–410. [CrossRef]
32. Helander, I.M.; Mattila-Sandholm, T. Fluorometric assessment of gram-negative bacterial permeabilization. *J. Appl. Microbiol.* **2000**, *88*, 213–219. [CrossRef]
33. Chojnacki, M.; Cao, X.; Young, M.; Fritz, R.N.; Dunman, P.M.; Flaherty, D.P. Optimization of 4-Substituted Benzenesulfonamide Scaffold To Reverse *Acinetobacter baumannii* Serum-Adaptive Efflux Associated Antibiotic Tolerance. *ChemMedChem* **2020**, *15*, 1731–1740. [CrossRef] [PubMed]
34. Kim, D.; Paggi, J.M.; Park, C.; Bennett, C.; Salzberg, S.L. Graph-based genome alignment and genotyping with HISAT2 and HISAT-genotype. *Nat. Biotechnol.* **2019**, *37*, 907–915. [CrossRef] [PubMed]
35. Perdigão Neto, L.V.; Oliveira, M.S.; Orsi, T.D.; Prado, G.; Martins, R.C.R.; Leite, G.C.; Marchi, A.P.; Lira, E.S.; Côrtes, M.F.; Espinoza, E.P.S.; et al. Alternative drugs against multiresistant Gram-negative bacteria. *J. Glob. Antimicrob. Resist.* **2020**, *23*, 33–37. [CrossRef]
36. Beganovic, M.; Daffinee, K.E.; Luther, M.K.; LaPlante, K.L. Minocycline Alone and in Combination with Polymyxin B, Meropenem, and Sulbactam against Carbapenem-Susceptible and -Resistant *Acinetobacter baumannii* in an In Vitro Pharmacodynamic Model. *Antimicrob. Agents Chemother.* **2021**, *65*, e01680-20. [CrossRef]
37. Singh, S.; Khanna, D.; Kalra, S. Minocycline and Doxycycline: More Than Antibiotics. *Curr. Mol. Pharmacol.* **2021**, *14*, 1046–1065. [CrossRef] [PubMed]
38. Seok, H.; Choi, W.S.; Lee, S.; Moon, C.; Park, D.W.; Song, J.Y.; Cheong, H.J.; Kim, J.; Kim, J.Y.; Park, M.N.; et al. What is the optimal antibiotic treatment strategy for carbapenem-resistant *Acinetobacter baumannii* (CRAB)? A multicentre study in Korea. *J. Glob. Antimicrob. Resist.* **2021**, *24*, 429–439. [CrossRef]
39. Gilani, M.; Latif, M.; Gilani, M.; Saad, N.; Ansari, M.; Gilani, M.; Waseem, H.; Naeem, A. Efficacy of Antimicrobials Against Multidrug-Resistant *Acinetobacter baumannii* from Patients in a Tertiary Care Hospital. *Microb. Drug Resist.* **2020**, *26*, 681–684. [CrossRef]
40. Ku, N.S.; Lee, S.H.; Lim, Y.S.; Choi, H.; Ahn, J.Y.; Jeong, S.J.; Shin, S.J.; Choi, J.Y.; Choi, Y.H.; Yeom, J.S.; et al. In vivo efficacy of combination of colistin with fosfomycin or minocycline in a mouse model of multidrug-resistant *Acinetobacter baumannii* pneumonia. *Sci. Rep.* **2019**, *9*, 17127. [CrossRef]
41. Flamm, R.K.; Shortridge, D.; Castanheira, M.; Sader, H.S.; Pfaller, M.A. In Vitro Activity of Minocycline against U.S. Isolates of *Acinetobacter baumannii*-Acinetobacter calcoaceticus Species Complex, *Stenotrophomonas maltophilia*, and *Burkholderia cepacia* Complex: Results from the SENTRY Antimicrobial Surveillance Program, 2014 to 2018. *Antimicrob. Agents Chemother.* **2019**, *63*, e01154-19. [CrossRef]

42. Dillon, N.; Holland, M.; Tsunemoto, H.; Hancock, B.; Cornax, I.; Pogliano, J.; Sakoulas, G.; Nizet, V. Surprising synergy of dual translation inhibition vs. *Acinetobacter baumannii* and other multidrug-resistant bacterial pathogens. *EBioMedicine* **2019**, *46*, 193–201. [CrossRef]
43. Fragkou, P.C.; Poulakou, G.; Blizou, A.; Blizou, M.; Rapti, V.; Karageorgopoulos, D.E.; Koulenti, D.; Papadopoulos, A.; Matthaiou, D.K.; Tsiodras, S. The Role of Minocycline in the Treatment of Nosocomial Infections Caused by Multidrug, Extensively Drug and Pandrug Resistant *Acinetobacter baumannii*: A Systematic Review of Clinical Evidence. *Microorganisms* **2019**, *7*, 159. [CrossRef] [PubMed]
44. Gardner, K.J.; Eady, E.A.; Cove, J.H.; Taylor, J.P.; Cunliffe, W.J. Comparison of serum antibiotic levels in acne patients receiving the standard or a modified release formulation of minocycline hydrochloride. *Clin. Exp. Dermatol.* **1997**, *22*, 72–76. [CrossRef] [PubMed]
45. Liu, F.; Rehmani, I.; Esaki, S.; Fu, R.; Chen, L.; de Serrano, V.; Liu, A. Pirin is an iron-dependent redox regulator of NF-κB. *Proc. Natl. Acad. Sci. USA* **2013**, *110*, 9722–9727. [CrossRef] [PubMed]
46. Orozco-Nunnelly, D.A.; Muhammad, D.; Mezzich, R.; Lee, B.S.; Jayathilaka, L.; Kaufman, L.S.; Warpeha, K.M. Pirin1 (PRN1) is a multifunctional protein that regulates quercetin, and impacts specific light and UV responses in the seed-to-seedling transition of *Arabidopsis thaliana*. *PLoS ONE* **2014**, *9*, e93371. [CrossRef]
47. Midgley, M. The phosphonium ion efflux system of *Escherichia coli*: Relationship to the ethidium efflux system and energetic studies. *J. Gen. Microbiol.* **1986**, *132*, 3187–3193. [CrossRef]
48. Stone, M.R.L.; Masi, M.; Phetsang, W.; Pagès, J.M.; Cooper, M.A.; Blaskovich, M.A.T. Fluoroquinolone-derived fluorescent probes for studies of bacterial penetration and efflux. *Medchemcomm* **2019**, *10*, 901–906. [CrossRef]
49. Fu, J.J.; Huang, D.Q.; Bai, Y.H.; Shen, Y.Y.; Lin, X.Z.; Huang, Y.; Ling, Y.R.; Fan, N.S.; Jin, R.C. How anammox process resists the multi-antibiotic stress: Resistance gene accumulation and microbial community evolution. *Sci. Total Environ.* **2022**, *807*, 150784. [CrossRef]
50. Yamada, J.; Yamasaki, S.; Hirakawa, H.; Hayashi-Nishino, M.; Yamaguchi, A.; Nishino, K. Impact of the RNA chaperone Hfq on multidrug resistance in *Escherichia coli*. *J. Antimicrob. Chemother.* **2010**, *65*, 853–858. [CrossRef]
51. Deter, H.S.; Abualrahi, A.H.; Jadhav, P.; Schweer, E.K.; Ogle, C.T.; Butzin, N.C. Proteolytic Queues at ClpXP Increase Antibiotic Tolerance. *ACS Synth. Biol.* **2020**, *9*, 95–103. [CrossRef]
52. Dechend, R.; Hirano, F.; Lehmann, K.; Heissmeyer, V.; Ansieau, S.; Wulczyn, F.G.; Scheidereit, C.; Leutz, A. The Bcl-3 oncoprotein acts as a bridging factor between NF-kappaB/Rel and nuclear co-regulators. *Oncogene* **1999**, *18*, 3316–3323. [CrossRef]
53. Wendler, W.M.; Kremmer, E.; Förster, R.; Winnacker, E.L. Identification of pirin, a novel highly conserved nuclear protein. *J. Biol. Chem.* **1997**, *272*, 8482–8489. [CrossRef] [PubMed]
54. Pang, H.; Bartlam, M.; Zeng, Q.; Miyatake, H.; Hisano, T.; Miki, K.; Wong, L.L.; Gao, G.F.; Rao, Z. Crystal structure of human pirin: An iron-binding nuclear protein and transcription cofactor. *J. Biol. Chem.* **2004**, *279*, 1491–1498. [CrossRef] [PubMed]
55. Adams, M.; Jia, Z. Structural and biochemical analysis reveal pirins to possess quercetinase activity. *J. Biol. Chem.* **2005**, *280*, 28675–28682. [CrossRef] [PubMed]
56. Gurmu, D.; Lu, J.; Johnson, K.A.; Nordlund, P.; Holmgren, A.; Erlandsen, H. The crystal structure of the protein YhaK from *Escherichia coli* reveals a new subclass of redox sensitive enterobacterial bicupins. *Proteins* **2009**, *74*, 18–31. [CrossRef]

antibiotics

Article

The RND Efflux Pump Gene Expression in the Biofilm Formation of *Acinetobacter baumannii*

Ola A. Abd El-Rahman [1,†], Fatma Rasslan [1,†], Safaa S. Hassan [2], Hossam M. Ashour [3,*] and Reham Wasfi [4]

1 Department of Microbiology and Immunology, Faculty of Pharmacy (Girls), Al-Azhar University, Cairo 11751, Egypt
2 Department of Clinical Pathology, National Cancer Institute, Cairo University, Cairo 11796, Egypt
3 Department of Integrative Biology, College of Arts and Sciences, University of South Florida, St. Petersburg, FL 33701, USA
4 Department of Microbiology and Immunology, Faculty of Pharmacy, October University for Modern Sciences and Arts (MSA), Giza 12451, Egypt
* Correspondence: ashour@usf.edu
† These authors contributed equally to this work.

Abstract: Multidrug resistant (MDR) *Acinetobacter baumannii* is a critical opportunistic pathogen in healthcare-associated infections (HAI). This is attributed to several factors, including its ability to develop biofilms that can enhance antimicrobial resistance (AMR) in addition to creating an environment for horizontal transfer of antibiotic resistance genes. The role of the efflux pump in biofilm formation is important for studies on alternative treatments for biofilms. One of the significant efflux pump families is the RND efflux pump family, which is common in Gram negative bacteria. The aim is to study the role of the RND efflux pump in biofilm formation by *A. baumannii*. The biofilm formation potential of thirty-four MDR *A. baumannii* isolates was evaluated by crystal violet assays. The effect of efflux pump inhibition and activation was studied using the efflux pump inhibitor carbonyl cyanide 3-chlorophenylhydrazone (CCCP) and the RND efflux pump substrate levofloxacin (at sub-MIC), respectively. The isolates were genotypically grouped by enterobacterial repetitive intergenic consensus (ERIC) typing and the expression of *adeABC*, *adeFGH*, and *adeIJK* efflux pump genes was measured by qPCR. Overall, 88.2% (30/34) of isolates were biofilm producers (the phenotype was variable including strong and weak producers). Efflux pump inhibition by CCCP reduced the biofilm formation significantly ($p < 0.05$) in 17.6% (6/34) of some isolates, whereas sub-MICs of the substrate levofloxacin increased biofilm formation in 20.5% (7/34) of other isolates. Overexpression of the three RND efflux pump genes was detected in five out of eleven selected isolates for qPCR with remarkable overexpression in the *adeJ* gene. No correlation was detected between the biofilm phenotype pattern and the RND efflux pump gene expression in biofilm cells relative to planktonic cells. In conclusion, the role of the RND efflux pumps AdeABC, AdeFGH, and AdeIJK in biofilm formation does not appear to be pivotal and the expression differs according to the genetic background of each strain. Thus, these pumps may not be a promising target for biofilm inhibition.

Keywords: *Acinetobacter baumannii*; biofilm; RND efflux pump; AdeABC; AdeFGH; AdeIJK

1. Introduction

Acinetobacter baumannii (*A. baumannii*) has emerged globally as an opportunistic multidrug-resistant (MDR) pathogen that is primarily linked with healthcare-associated infections, including pneumonia, meningitis, bacteremia and urinary tract infections [1]. *A. baumannii* is characterized by its natural competence that makes it more capable of incorporating foreign DNA, which can subsequently increase its adaptability to hostile environments [2,3]. The emergence of resistant isolates to widely prescribed antimicrobial agents leads to a high rate of treatment failure, and a higher mortality rate for infections caused by *Acinetobacter* spp. The poor prognosis of infections caused by *A. baumannii* puts this bacterium among the list of "top critical threat" pathogens [4]. Various mechanisms are associated with the multidrug resistance phenotype, including enzymatic degradation, alteration of a target site, decreased membrane permeability, biofilm formation, and increased expression of the efflux pumps [5]. It was reported that the efflux pumps play a dual role in antibiotic resistance either directly by extruding antibiotics or indirectly by biofilm formation. Hence, the correlation between the wide substrate specificity efflux pump systems and the bacterial ability to form biofilms has become an important area to explore [6].

The efflux pumps implicated in bacterial multidrug resistance perform a variety of roles in the transition of planktonic cells to biofilms. These pumps actively extrude quorum-sensing auto-inducers, as well as antimicrobial agents and metabolic intermediates, resulting in direct and indirect regulation of biofilm formation and quorum sensing [7]. Some pumps also cause change in cell membrane composition, thus, altering biofilm formation [8].

A. baumannii are characterized by six major efflux pump families, which are major facilitator superfamily (MFS), multidrug and toxic efflux (MATE), resistance nodulation-division (RND), small multidrug resistance (SMR), ATP binding cassette (ABC) [9], and the novel proteobacterial antimicrobial compound efflux (PACE) family [10]. Among the efflux systems, RND pumps are the most prevalent systems in Gram-negative bacteria [11]. RND efflux system extrudes a wide range of substrates that are structurally unrelated [12]. Three *Acinetobacter* drug efflux pumps (Ade) belong to the RND systems, namely AdeABC, AdeFGH, and AdeIJK, which have been widely distributed in *A. baumannii* [13]. In addition, these efflux pumps have common substrates, which are fluoroquinolones and chloramphenicol [11]. Little is known regarding the association between efflux pumps in *A. baumannii* and biofilm formation, hence, the aim of our study is to investigate the role of the RND efflux pump system in biofilm formation by *A. baumannii*. This can open new avenues for the treatment of biofilm-related infections.

2. Results

2.1. Biofilm Formation Ability of A. baumannii Isolates and Enterobacterial Repetitive Intergenic Consensus (ERIC) Genotyping

Thirty-four MDR *A. baumannii* isolates (collected and genotyped by ERIC typing) were grouped into 9 ERIC clusters (A to I). The biofilm formation index (BFI) of the tested isolates was classified as strong, moderate, weak, and non-adherent, as in Table 1. Thus, eight of the isolates were classified as strong (8/34; 23.5%), ten as moderate (11/34; 32.3%), eleven as weak (11/34; 32.3%), and five as non-adherent (4/34; 11.7%) (Table 1). The standard strain showed strong biofilm forming ability (BFI = 1.7). Each ERIC cluster was found to include isolates with the same biofilm phenotype except clusters G and H, which included isolates with different biofilm phenotypes (Table 1).

Table 1. Biofilm formation phenotypes of *Acinetobacter baumannii* isolates genotypically characterized by ERIC-PCR.

ERIC Cluster Group	Isolate Number	Biofilm Formation Pattern
A	25	Strong
	30	Strong
B	7	Weak
	9	Weak
	37	Weak
C	22	Strong
	27	Moderate
	39	Strong
	43	Strong

Table 1. *Cont.*

ERIC Cluster Group	Isolate Number	Biofilm Formation Pattern
D	1	Moderate
	26	Moderate
	44	Moderate
	45	Moderate
	47	Moderate
E	32	Moderate
	33	Moderate
	46	Moderate
F	31	Moderate
	35	Moderate
G	3	Weak
	6	Weak
	24	Non-Adherent
	28	Strong
	29	Weak
	36	Non-Adherent
	40	Non-Adherent
	41	Non-Adherent
H	2	Weak
	4	Strong
	14	Weak
	15	Weak
	23	Strong
I	12	Weak
	21	Weak

2.2. *The Effect of Efflux Pump Substrate and Efflux Pump Inhibitor on Biofilm Phenotype by A. baumannii Isolates*

2.2.1. Efflux Pump Substrate (Levofloxacin)

Levofloxacin-resistant isolates represented (25/34; 73.5%) of the isolates as confirmed by the high MICs of levofloxacin ranging from 4 µg/mL to 216 µg/mL. In order to determine the role of the efflux pump in levofloxacin resistance, CCCP was used at a concentration of 12.5 µg/mL, which does not reduce the growth of isolates as reflected by their OD_{600}, compared to control untreated culture. The addition of CCCP to the bacterial culture of levofloxacin-resistant isolates caused a reduction in MIC by four folds or more in 56% (14/25) of resistant isolates as shown in Table 2. On the other hand, 40% (10/25) of levofloxacin-resistant isolates became sensitive upon treatment with CCCP (Table 2).

Table 2. Change in the phenotypic resistance pattern of isolates to levofloxacin under the effect of the efflux pump inhibitor carbonyl cyanide 3-chlorophenylhydrazone (CCCP) (*n* = 25).

Isolate Number	MIC (µg/mL)	MIC + CCCP (µg/mL)	Phenotypic Factor (F) # CCCP	Resistance Phenotype † (−/+ CCCP)
A. baumannii ATCC 19606	1	1	1	S/S
4	64	32	2	R/R
7	8	4	2	R/R
12	8	4	2	R/R
14	64	32	2	R/R
15	8	4	2	R/R
22	4	1	4 *	R/S
23	8	1	8 *	R/S
24	64	1	64 *	R/S
25	8	1	8 *	R/S
26	64	1	64 *	R/S
27	4	1	4 *	R/S
28	256	128	2	R/R
29	256	128	2	R/R

Table 2. *Cont.*

Isolate Number	MIC (μg/mL)	MIC + CCCP (μg/mL)	Phenotypic Factor (F) # CCCP	Resistance Phenotype † (−/+ CCCP)
30	256	1	256 *	R/S
32	64	8	8 *	R/R
33	16	4	4 *	R/R
35	32	8	4 *	R/R
36	32	1	32 *	R/S
37	32	16	2	R/R
39	8	1	8 *	R/S
40	64	32	2	R/R
41	128	32	4 *	R/R
43	4	2	2	R/I
44	32	16	2	R/R
45	16	2	8 *	R/I

† S: Susceptible (MIC ≤ 1 μg/mL), I: Intermediate (MIC = 2 μg/mL), R: Resistant (MIC ≥ 4μg/mL). # Phenotypic efflux factor (F) = MIC without CCCP (EPI)/MIC with CCCP (EPI); * Isolates showing > 4 phenotypic efflux factor with CCCP 12.5 μg/mL.

Table 3. Biofilm formation potential by *Acinetobacter baumannii* isolates and the effect of efflux pump inhibitor (CCCP) and efflux pump substrate (levofloxacin at 0.25x MIC) on biofilm formation.

ERIC Cluster Group	Isolate Number	BFI #	Biofilm Formation Pattern	BFI with Sub-MIC of Levofloxacin (Fold Change)	BFI with CCCP (Fold Change)
A	25 #	2.60	Strong	2.30(0.91)	2.20 (0.88)
	30	1.30	Strong	1.24 (0.92)	1.04 (0.80)
B	7 #	0.70	Weak	0.95 (1.35) *	0.70 (1.00)
	9	0.70	Weak	0.73 (1.05)	0.80 (1.17)
	37	0.60	Weak	0.54 (0.90)	0.61 (1.02)
C	22	1.27	Strong	1.10 (0.90)	1.20 (0.94)
	27 #	2.37	Moderate	2.48 (1.05)	2.08 (0.88) *
	39	1.43	Strong	1.37 (0.89)	1.43 (1.00)
	43	1.10	Strong	0.99 (0.90)	1.10 (0.90) *
D	1	0.80	Moderate	0.84 (1.05)	0.90 (1.12)
	26	0.70	Moderate	0.88 (1.26) *	1.24 (1.77)
	44	0.90	Moderate	1.80 (2.00) **	0.85 (0.95)
	45 #	0.79	Moderate	0.86 (1.08)	0.66 (0.83) *
	47	0.90	Moderate	1.00 (1.10)	0.80 (0.88)
E	32	0.95	Moderate	0.90 (0.94)	0.95 (1.00)
	33 #	0.96	Moderate	0.86 (0.90)	0.96 (1.00)
	46	0.80	Moderate	0.90 (1.12)	0.85 (1.06)
F	31	0.90	Moderate	0.95 (1.05)	1.00 (1.10)
	35 #	1.68	Moderate	1.39 (0.83)	1.30 (0.78)
G	3	0.65	Weak	0.70 (1.07)	0.80 (1.23)
	6	0.60	Weak	0.65 (1.08)	0.58 (0.96)
	24	0.29	Non-Adherent	0.28 (0.99)	0.49 (1.70)
	28 #	1.50	Strong	1.48 (0.98)	1.29 (0.86) *
	29	0.62	Weak	0.56 (0.90)	0.62 (1.00)
	36 #	0.29	Non-Adherent	0.46 (1.30) *	0.29 (1.00)
	40	0.04	Non-Adherent	0.19 (5.00) *	0.035 (0.90)
	41	0.11	Non-Adherent	0.27 (2.50) **	0.09 (0.90)
H	2	0.60	Weak	0.64 (1.08)	0.67 (1.12)
	4 #	2.00	Strong	1.78 (0.89)	2.34 (1.17)
	14	0.55	Weak	0.48 (0.89)	0.42 (0.76)
	15	0.59	Weak	0.87 (1.49) *	0.67 (0.40) *
	23 #	2.40	Strong	2.16 (0.90)	0.79 (0.33) *
I	12 #	0.52	Weak	0.58 (1.13)	0.42 (0.80)
	21	0.50	Weak	0.60 (1.20)	0.55 (1.10)

BFI: ≥1.1 = strong, 0.7–1.09 = moderate, 0.35–0.69 = weak, <0.35 = non adherent. # Isolates selected for gene expression analysis by qPCR. Statistical analysis by unpaired t-test, with significance * $p < 0.05$, ** $p < 0.01$.

After determining the MIC of levofloxacin against isolates, the effect of increased efflux pump activity by sub-MICs of levofloxacin on biofilm formation was studied. The addition of sub-MICs of levofloxacin to bacterial culture significantly increased ($p < 0.05$) the biofilm formation pattern of 20.5% (7/34) of the isolates (Table 3). No correlation was detected between fold change in the BFI using sub-MIC of levofloxacin and the biofilm formation potential of the isolates (Spearman's $r_s = -0.47, p > 0.05$).

2.2.2. Efflux Pump Inhibitor (CCCP)

Using CCCP for efflux pump inhibition significantly reduced ($p < 0.05$) biofilm formation in 17.6% (6/34) of the isolates (Table 3). However, isolates showing change in biofilm formation potential in the presence of CCCP were different from those affected by sub-MICs of levofloxacin. No correlation was detected between fold change in BFI using CCCP and the biofilm formation potential of the isolates (Spearman's $r_s = 0.07, p > 0.05$).

2.3. Expression Levels of the adeB, adeG, and adeJ Efflux Pump Genes in Planktonic and Biofilm Cells Using Quantitative Real-Time PCR (qRT-PCR)

The three efflux genes were detected by conventional PCR in the thirty-four isolates and the standard *A. baumannii* strain (ATCC 19606). Gene expression analysis was done on eleven isolates and a standard strain by qRT-PCR analysis to measure the expression of the three RND efflux pumps. Isolates for gene expression were selected based on the ERIC cluster typing by selecting an isolate from each cluster that represents the biofilm phenotype in this group. A representative isolate was selected from each ERIC cluster except for clusters G and H, which comprise two biofilm phenotypes, therefore, two isolates were selected from these groups. Measuring the expression levels of *adeB*, *adeG*, and *adeJ* genes showed overexpression in five out of the eleven tested isolates relative to the standard strain (Figure 1a). Isolates with overexpression in the efflux pump were moderate (isolates no. 27, 33, 35, 45) and weak biofilm forming (isolate no. 7). It was observed that the *adeJ* gene relative expression was remarkably higher than the other two RND efflux pump genes in tested isolates.

The relative expression pattern of *adeB*, *adeG*, and *adeJ* varied among different isolates with different biofilm forming abilities. Five strong biofilm forming isolates (Standard strain ATCC 19606, 4, 23, 25, and 28) showed more than a twofold increase in the gene expression of efflux pumps in biofilm form than planktonic form, while the remaining strong and moderate biofilm isolates (27, 33, 35, 45) showed either more than a twofold reduction or no change in gene expression of the same efflux pumps compared to planktonic forms (Figure 1b). No correlation was detected between the biofilm formation ability of isolates and the relative expression of the RND efflux pump genes (Spearman's $r_s = 0.17$, 0.14, and 0.35 for *adeB*, *adeG*, and *adeJ*, respectively; $p > 0.05$). Isolates that belong to the same ERIC clusters (G:28–36 and H:4–23) or closely related clusters (A:25 and B:7) showed similar efflux pump relative expression patterns despite the difference in biofilm phenotype in some cases.

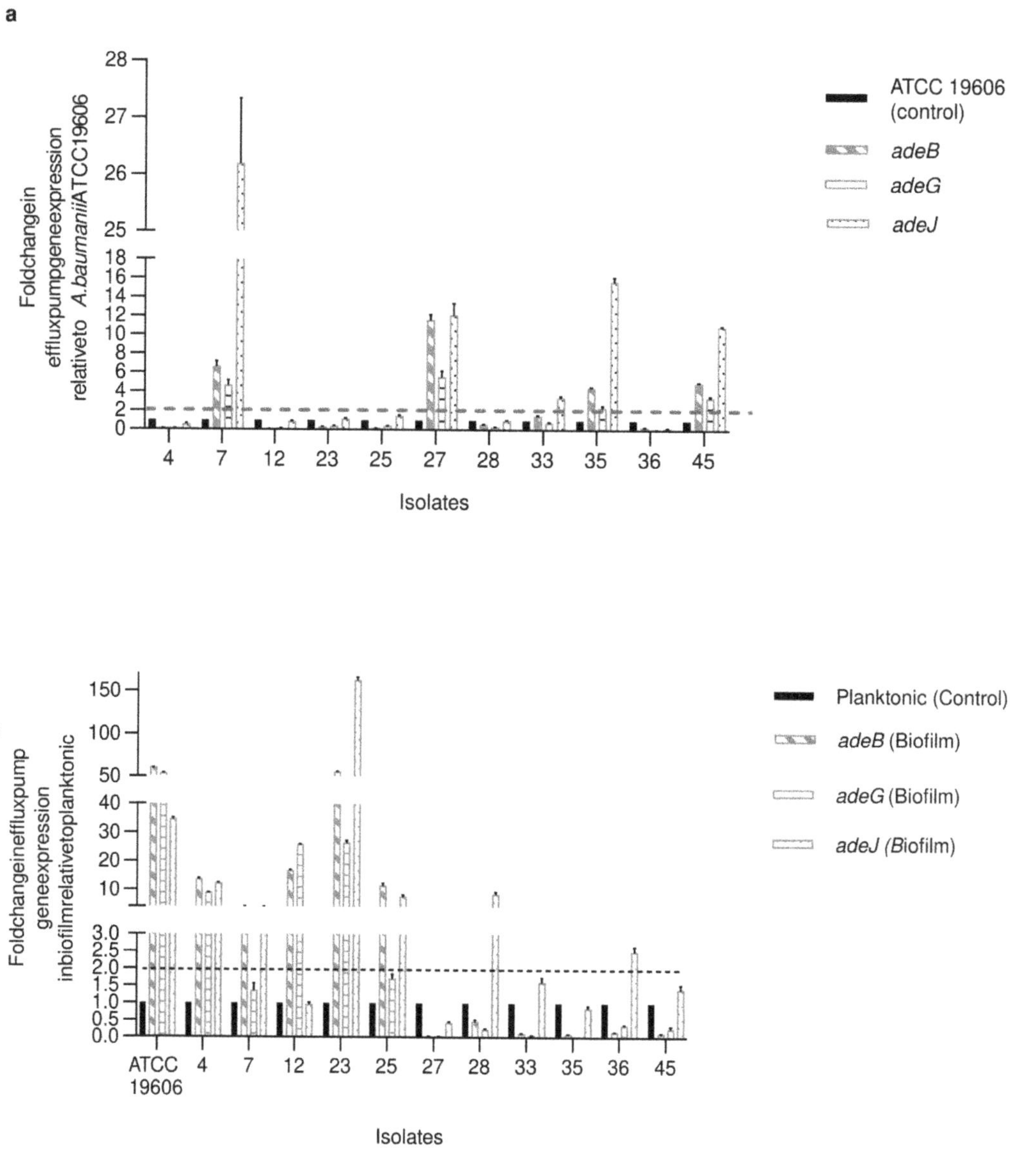

Figure 1. Fold changes in expression of the *adeB*, *adeG*, and *adeJ* genes in *A. baumannii* isolates. The bars represent relative expression levels of each gene with the corresponding control. (**a**) Expression of isolates in planktonic cells relative to standard strain (ATCC 19606) (control); (**b**) Expression of isolates in biofilm forming cells and standard strain relative to their planktonic cells (control). The error bars represent the standard error of mean (SEM). All values are means ± SEM. 16S rRNA was used as housekeeping gene control. Strong biofilm forming isolate (Standard strain ATCC 19606, 4, 23, 25, 28), moderate biofilm forming isolate (27, 33, 35, 45), weak biofilm forming (7, 12), and non-adherent isolate (36). Dotted line: Change in expression > 2 folds.

3. Discussion

A. baumannii has become a major therapeutic problem worldwide due to its high prevalence in hospital environments and acquisition of antimicrobial resistance [6]. *A. baumannii* has developed different mechanisms to resist antibiotics and among them is extrusion of toxic compounds by efflux pumps and biofilm formation [14]. The resistance nodulation division (RND) efflux pump family represents the major type of efflux pump in Gram negative bacteria [15]. Therefore, it has been the focus of our current study. Evaluating the biofilm formation potential of the MDR *A. baumannii* isolates in our study revealed that, overall, 88.2% (30/34) of isolates were biofilm producers with variable phenotypes that ranged from strong to weak. The presence of weak and non-adherent biofilm isolates representing 44.1% of the MDR isolates indicates that biofilm formation has no association with antibiotic resistance as reported by Chen and colleagues [13].

Recently, the role of efflux pumps in biofilm formation has been the focus of different studies seeking alternatives to antimicrobials in controlling biofilm formation [6,16]. To study the effect of the efflux pump substrate and efflux pump inhibitors on biofilm phenotypes, we utilized sub-MICs of levofloxacin and CCCP, respectively. Levofloxacin is a common substrate for the three RND efflux pumps: AdeABC, AdeFGH, and AdeIJK [17,18].

In the present study, the resistance of *A. baumannii* isolates to levofloxacin was detected in 73.5% (25/34) of the isolates, a percentage that is comparable to findings of a previous study by Zaki and colleagues [19] but less than resistance percentages detected in other studies that showed resistance rates of 85–92% [20,21]. The susceptibility of 56% (14/25) of strains to levofloxacin has increased in the presence of the efflux pump inhibitor CCCP; however, only 40% of resistant isolates have restored susceptibility under the effect of CCCP. Similar reduction in quinolone resistance after the addition of CCCP was reported in previous studies [22,23], indicating the involvement of efflux system in these isolates. The failure of CCCP in restoring susceptibility to quinolone in some isolates can indicate the presence of more than one mechanism of resistance to levofloxacin in these isolates, such as mutations in the *gyrA* or *parC* genes encoding gyrase subunit A and topoisomerase IV subunit C, respectively [24]. The addition of sub-MICs of levofloxacin (efflux pump substrate) to bacterial culture significantly increased the biofilm formation pattern ($p < 0.05$) in 20.5% (7/34) of isolates. A study by He and colleagues [25] illustrated that sub-MIC of levofloxacin caused overexpression of the *adeFGH* gene, resulting in accelerated synthesis and transport of quorum sensing molecules during biofilm formation. On the other hand, efflux pump inhibition by CCCP significantly reduced ($p < 0.05$) biofilm formation in 17.6% (6/34) of isolates. Strains affected by levofloxacin were different from those affected by CCCP indicating that the efflux pumps impacted in each of the two cases were different [26]. Moreover, the exposure of *A. baumannii* to sub-inhibitory concentrations of antimicrobials did not only increase substrate efflux by the efflux pump but also changed the expression of genes controlling biofilm formation, including genes regulating pili, other efflux pumps, and virulence factors [27].

The RND efflux pump genes were detected in all tested isolates. Likewise, in another study, AdeABC, AdeFGH, and AdeIJK pumps were detected in 92.18%, 98.43%, and 89.06% of strains, respectively [28]. The relative gene expression of the major part of tripartite efflux systems (*adeB*, *adeG*, and *adeJ*) was measured by qPCR for eleven isolates representing different ERIC clusters relative to that of the reference *A. baumannii* strain (ATCC 19606). CCCP increased the levofloxacin susceptibility of the five isolates that did not show overexpression for any of the efflux pumps. No correlation was detected between the effect of CCCP on resistance to levofloxacin as indicated by a reduction in the MIC and the RND efflux pump expression (Spearman's $r_s = -0.08$, -0.16, and -0.02, for *adeB*, *adeG*, and *adeJ*, respectively; $p > 0.05$). This result indicates the involvement of other efflux pumps in levofloxacin resistance by these isolates, such as *abeM* and *abaQ*, that belong to the MATE and MFS efflux systems, respectively [29]. The previous efflux pumps can also be inhibited by CCCP [30,31].

Previous studies on the role of *A. baumannii* efflux pumps in biofilm formation were based on knocking out the RND efflux pump genes in a limited number of strains and then measuring the impact of this on biofilm formation [16,32,33] or on measuring the relative expression of efflux genes in the planktonic form and correlating the expression levels to the degree of biofilm formation [6,13]. In our assessment, these previous approaches are not adequate or even representative for the bacteria, given its distinct phenotypic and genotypic characteristics. Moreover, these methods do not reflect the actual expression pattern in the biofilm form. Thus, the expression of *adeB, adeG,* and *adeJ* efflux pump genes in the biofilm form relative to the planktonic form was studied to illustrate the role of these pumps in the mature biofilm of *A. baumannii*. Results have shown that the relative expression pattern of the RND pumps genes varied among different isolates with different biofilm forming abilities. The lack of correlation between efflux pump gene expression and biofilm formation potential suggests that the RND efflux pump role in biofilm formation is not pivotal. Some studies reported the lack of association between the activity of the efflux pump family and biofilm formation [6,13,28]. Other studies reported a negative correlation between the activity of one or more of the RND efflux pump genes *adeB, adeG,* and *adeJ* and biofilm formation [6,12,16,32,34–36]. The negative correlation observation was explained by the presence of associated changes in the bacterial membrane with efflux pump overexpression, such as the reduction in *Csu-pili* and the downregulation of genes involved in iron acquisition and motility [6,16,33]. Furthermore, other studies reported that the increase in expression of one or more of the *adeB, adeG,* and *adeJ* genes was associated with increased biofilm formation [6,16,25,28,33,36]. The overexpression of the *adeFGH* efflux pump gene was suggested to play a role in the synthesis and efflux of quorum sensing molecules during biofilm formation [25,37].

Other explanations highlighted operons as the controllers of these pumps through controlling other genes contributing to biofilm formation. In this scenario, the efflux pumps are not directly in control of the biofilm phenotype. For example, the low expression of *adeABC* is associated with the reduced expression of other genes involved in biofilm formation, motility, and virulence because these genes are subjected to control by the regulator adeRS [33]. The global regulator *adeN* of the *adeIJK* does not only regulate the *adeIJK* expression but also the expression of many other factors involved in biofilm formation [38]. In addition, the levels of guanosine tetraphosphate (ppGpp) could control the expression of *adeABC* and *adeIJK* [39,40]. Another controller for the expression of the *adeIJK* and *adeABC* efflux pump genes is the BaeSR (two component system), which is involved in controlling the expression of other efflux pumps that favor biofilm formation (such as MacAB-TolC) through cell detoxification and maintenance responses [39,41].

In previous studies, different *A. baumannii* strains exhibited different responses to RND efflux pump gene deletion and the role of these pumps in biofilm formation may differ between the strains [6,16,33]. *A. baumannii* was shown to require certain expression profiles of efflux pumps to initiate and maintain biofilm formation [33]. Furthermore, the overexpression profile of the RND pumps could be specific for each country [8]. In our study, it was observed that the *adeJ* gene relative expression was remarkably higher than the two other RND efflux pump genes in tested isolates. Similarly, in French and Korean studies, overexpression of the *adeJ* gene was detected as compared to the expression of other RND pumps [6,29]. Overexpression of *adeG* was reported in isolates from Canada [42].

Amin and colleagues reported that isolates that are genotypically characterized according to ERIC typing showed similar biofilm formation potential [28]. Similarly, we observed that isolates that belong to the same ERIC clone showed similar biofilm forming potential, except in clusters B and G. Notwithstanding, isolates that belong to the same ERIC clusters (G:28–36 and H:4–23) or closely related clusters (A:25 and B:7) showed similar efflux pump relative expression patterns in biofilm form compared to the planktonic form despite the difference in the biofilm phenotype.

In conclusion, overexpression of the RND efflux pumps AdeABC, AdeFGH, and AdeIJK was not correlated with the biofilm phenotype in any of the tested isolates. Therefore, their roles in biofilm formation do not appear to be pivotal and their expression differs according to the genetic background of each strain. Consequently, efflux pumps may be a part of more complicated mechanisms that contribute to biofilm formation.

4. Materials and Methods

4.1. Bacterial Isolates

Thirty-four Multidrug Resistant (MDR) *A. baumannii* isolates from blood samples of cancer patients at the National Cancer Institute (NCI) from our previous study were genotypically characterized via enterobacterial repetitive intergenic consensus (ERIC)–PCR [2]. Approval was obtained from the ethics committee of the October University for Modern Sciences and Arts. Standard *A. baumannii* (ATCC 19606) was used as a control for the gene expression assay.

4.2. Assessment of the Biofilm Forming Ability of Isolates

Crystal violet assays were used to determine the biofilm forming ability of the isolates [43]. The turbidity of an overnight culture was adjusted to the equivalent of McFarland 0.5, and 200 µL of the adjusted culture was added to each well of a 96-well microtiter plate. After overnight incubation at 37 °C, growth turbidity was measured at 600 nm (OD_{growth}), then the biofilm was stained by crystal violet (2% *w*/*v*). Plates were left for air drying before destaining with glacial acetic acid (33% *v*/*v*). Finally, biofilm biomass was measured using a microplate reader at 545 nm (OD_{CV}). Biofilm formation was evaluated using a biofilm formation index [BFI]: (OD_{CV} Biofilm-OD_{CV} Control)/OD_{growth}. The biofilm formation ability of the isolates was classified into strong, moderate, weak, and non-adherent according to the BFI values described by Yaikhan and colleagues [44].

4.3. Evaluation of the Role of Efflux Pumps on Biofilm Formation Phenotypes

The efflux pump inhibitor (EPI) named carbonyl cyanide 3-chlorophenylhydrazone (CCCP) (ChemCruz, Dallas, Texas, USA) was used to phenotypically evaluate the role of efflux pumps in biofilm formation and levofloxacin resistance. The effects of two commonly used concentrations of CCCP (12.5 µg/mL and 25 µg/mL) were tested on bacterial growth. The concentrations that did not reduce growth significantly as determined spectrophotometrically at OD_{600}, compared to control untreated culture, were selected for efflux pump inhibition tests [29].

Efflux pump substrate was used to study the effect of efflux pump activation on biofilm formation. Levofloxacin is a common substrate for the three RND efflux pumps under study, and the role of this substrate on the pump activity was studied at sub-MIC levels. Firstly, the MICs of levofloxacin were determined in addition to the change in MICs in the presence of CCCP. A calculated phenotypic efflux factor F (MIC without CCCP (EPI) /MIC with CCCP (EPI)) of 4 and above reflects the role of efflux pumps in levofloxacin resistance [45]. Secondly, the effect of efflux pump on biofilm phenotype was determined by assessing the difference in BFI in presence of CCCP and sub-MICs of levofloxacin which are used as EPI and efflux pump substrate, respectively. Fold change in biofilm = BFI of untreated isolate/BFI in presence of CCCP or sub-MIC of levofloxacin [25,46].

4.4. Detection of the Efflux Pump Genes *adeABC*, *adeFGH*, and *adeIJK*

Pure colonies of *A. baumannii* isolates were used for DNA extraction using the Thermo Scientific™ GeneJet™ genomic DNA purification kit (Thermo Scientific, Waltham, MA, USA), in accordance with the manufacturer's recommendations, and then kept at −20 °C. The genes *adeB*, *adeG*, and *adeJ* were selected for amplification. These genes were selected because they represent an essential part of the Ade efflux system. PCR was performed to screen for the presence of the efflux transporter genes *AdeB*, *AdeG*, and *AdeJ*, using the primers and annealing temperatures described in Table 4. Genes were amplified by initial denaturation at 94 °C for 5 min, followed by 35 cycles consisting of 3 phases: DNA denaturation at 94 °C for 0.5 min, annealing according to Table 4 for 0.5 min, elongation at 72 °C for 0.5 min, and a final elongation at 72 °C for 10 min.

Table 4. Oligonucleotides sequences and annealing temperatures.

Primer	Sequence 5′—3′	Annealing Temperature (°C)	Ampilicon Size bp	Reference
adeB	F: GCAGAGCGTACTCGGAATGT R:CCACTGAAACCCCATCCCAA	57	101	[47]
adeG	F:GCGTTGCTGTGACAGATGTT R:TTGTGCACGGACCTGATAAA	52	104	[5]
adeJ	F:TTCGGTGGCTCATACGCAAT R:GGAGCACCACCTAACTGACC	57	137	[47]
16s rRNA	F:AGCTAACGCGATAAGTAGACCG R:TGTCAAGGCCAGGTAAGGTTC	57	137	[47]

4.5. Extraction of Total Bacterial RNA from Planktonic and Biofilm Forming Cells

A representative isolate from each ERIC cluster was selected for gene expression analysis by the Quantitative polymerase chain reaction (qPCR). Isolates were grown overnight at 37 °C in Lauria Bertani (LB) broth and diluted to McFarland 0.5. A volume of 7 mL of adjusted suspension from each isolate was added in each well of 6-well plates (Greiner Bio-One, Kremsmünster, Austria) and was incubated at 37 °C for 24 h. After incubation, culture suspension was removed from wells for extracting RNA from planktonic bacteria. Cells adhering to the plate wells were washed twice by cold sterile Phosphate buffer saline (PBS) and, then, dislodged by sonication before suspension in falcon tube containing PBS. Total RNA was isolated from planktonic and adherent cells using RNeasy® Minikit (Qiagen, Hilden, Germany) according to the manufacturer's instructions. Contaminating DNA was removed by RNase-free DNase I (New England Biolabs, USA). The concentrations and purity of RNA were quantified with a NanoDrop™ Spectrophotometer (ThermoScientific™, Ipswich, MA, USA) at 260 and 280 nm (260/280 ratio of >1.8) and agarose gel electrophoresis of RNA samples verified its integrity. Finally, the RevertAid First strand cDNA synthesis kit (ThermoScientific™, Waltham, MA, USA) was used to reverse transcribe 1 µg of total RNA sample into cDNA.

4.6. Quantification of the Expression of the RND Efflux Pump Genes (adeB, adeG, adeJ) in Biofilm and Planktonic Forms Using RT-qPCR

Real-time quantitative PCR was used to measure the relative expression of the RND-family efflux pump genes (*adeB, adeG, adeJ*) using Maxima SYBR Green qPCR Master Mix with ROX (ThermoScientific, Waltham, Massachusetts, USA) and primers listed in Table 4. The expression of pump genes was normalized to the housekeeping gene (16S rRNA). Expression quantification was done using the StepOnePlus™ Real-Time PCR System and software (Applied Biosystems, Waltham, Massachusetts, USA) according to the manufacturer's instructions. After a 10 min activation of the modified Taq polymerase at 95 °C, 40 cycles of 15 s at 95 °C, 30 s at 60 °C, and 30 s at 72 °C were performed. Data acquisition was done at 60 °C and 72 °C. The relative gene expression for pump genes transcripts was calculated against internal control 16S rRNA gene. The relative expression of efflux pump genes in the tested isolates was calculated against those of *A. baumannii* ATCC 19606 (expression = 1), which served as the control to determine overexpressed pump genes. The expression of the efflux pump genes in the biofilm form was determined relative to their expression in the planktonic form. The $2^{-\Delta\Delta CT}$ method was used to calculate the relative expression level of pump genes. A significant effect on gene expression was concluded when the corresponding ratios were >2.0. All reactions were performed in triplicate [24].

4.7. Statistical Analysis

We used the Prism version 8.3.0 software for Windows (GraphPad Software) for data analysis and graph plotting. The relationships between different factors were assessed by calculating Spearman's correlation coefficients. The student's t-test was used for data analysis of the gene expression. Statistical differences were considered significant at $p < 0.05$.

Author Contributions: Conceptualization, O.A.A.E.-R., F.R., H.M.A. and R.W.; Methodology, O.A.A.E.-R., F.R., H.M.A. and R.W.; Validation, O.A.A.E.-R., F.R., S.S.H., H.M.A. and R.W.; Formal analysis, O.A.A.E.-R., F.R., S.S.H., H.M.A. and R.W.; Investigation, O.A.A.E.-R., F.R. and R.W.; writing—original draft preparation, O.A.A.E.-R., F.R., H.M.A. and R.W.; Data curation; O.A.A.E.-R., F.R., S.S.H., H.M.A. and R.W.; writing—review and editing, O.A.A.E.-R., F.R., S.S.H., H.M.A.; Visualization, R.W.; Supervision, H.M.A. and R.W., Project administration, H.M.A. and R.W. All authors have read and agreed to the published version of the manuscript.

Funding: This research received no external funding.

Data Availability Statement: All relevant data are included within the manuscript.

Conflicts of Interest: The authors declare no conflict of interest.

References

1. Kim, H.-R.; Shin, D.-S.; Jang, H.-I.; Eom, Y.-B. Anti-biofilm and anti-virulence effects of zerumbone against *Acinetobacter baumannii*. *Microbiology* **2020**, *166*, 717–726. [CrossRef] [PubMed]
2. Wasfi, R.; Rasslan, F.; Hassan, S.S.; Ashour, H.M.; El-Rahman, O.A.A. Co-Existence of Carbapenemase-Encoding Genes in *Acinetobacter baumannii* from Cancer Patients. *Infect. Dis. Ther.* **2021**, *10*, 291–305. [CrossRef] [PubMed]
3. Zafer, M.M.; Hussein, A.F.A.; Al-Agamy, M.H.; Radwan, H.H.; Hamed, S.M. Genomic Characterization of Extensively Drug-Resistant NDM-Producing *Acinetobacter baumannii* Clinical Isolates With the Emergence of Novel blaADC-257. *Front. Microbiol.* **2021**, *12*, 736982. [CrossRef] [PubMed]
4. Kengkla, K.; Kongpakwattana, K.; Saokaew, S.; Apisarnthanarak, A.; Chaiyakunapruk, N. Comparative efficacy and safety of treatment options for MDR and XDR *Acinetobacter baumannii* infections: A systematic review and network meta-analysis. *J. Antimicrob. Chemother.* **2018**, *73*, 22–32. [CrossRef]
5. Nowak, J.; Seifert, H.; Higgins, P. Prevalence of eight resistance-nodulation-division efflux pump genes in epidemiologically characterized *Acinetobacter baumannii* of worldwide origin. *J. Med. Microbiol.* **2015**, *64*, 630–635. [CrossRef]
6. Kim, C.-M.; Park, G.; Ko, Y.J.; Kang, S.-H.; Jang, S.J. Relationships between Relative Expression of RND Efflux Pump Genes, H33342 Efflux Activity, Biofilm-Forming Activity, and Antimicrobial Resistance in *Acinetobacter baumannii* Clinical Isolates. *Jpn. J. Infect. Dis.* **2021**, *74*, 499–506. [CrossRef]
7. Navidifar, T.; Amin, M.; Rashno, M. Effects of sub-inhibitory concentrations of meropenem and tigecycline on the expression of genes regulating pili, efflux pumps and virulence factors involved in biofilm formation by *Acinetobacter baumannii*. *Infect. Drug Resist.* **2019**, *12*, 1099–1111. [CrossRef]
8. Kornelsen, V.; Kumar, A. Update on Multidrug Resistance Efflux Pumps in *Acinetobacter* spp. *Antimicrob. Agents Chemother.* **2021**, *65*, e00514-21. [CrossRef]
9. Abdi, S.N.; Ghotaslou, R.; Ganbarov, K.; Mobed, A.; Tanomand, A.; Yousefi, M.; Asgharzadeh, M.; Kafil, H.S. *Acinetobacter baumannii* Efflux Pumps and Antibiotic Resistance. *Infect. Drug Resist.* **2020**, *13*, 423–434. [CrossRef]
10. Hassan, K.A.; Liu, Q.; Elbourne, L.D.; Ahmad, I.; Sharples, D.; Naidu, V.; Chan, C.L.; Li, L.; Harborne, S.P.; Pokhrel, A.; et al. Pacing across the membrane: The novel PACE family of efflux pumps is widespread in Gram-negative pathogens. *Res. Microbiol.* **2018**, *169*, 450–454. [CrossRef]
11. Barnie, P.A.; Xing, L.; Su, Z.; Xu, H. Development of Efflux Pumps and Inhibitors (EPIs) in *A. baumanii*. *Clin. Microbiol. Open Access* **2014**, *3*, 1. [CrossRef]
12. Yoon, E.-J.; Chabane, Y.N.; Goussard, S.; Snesrud, E.; Courvalin, P.; Dé, E.; Grillot-Courvalin, C. Contribution of Resistance-Nodulation-Cell Division Efflux Systems to Antibiotic Resistance and Biofilm Formation in *Acinetobacter baumannii*. *Mbio* **2015**, *6*, e00309-15. [CrossRef] [PubMed]
13. Chen, L.; Li, H.; Wen, H.; Zhao, B.; Niu, Y.; Mo, Q.; Wu, Y. Biofilm formation in *Acinetobacter baumannii* was inhibited by PAβN while it had no association with antibiotic resistance. *Microbiologyopen* **2020**, *9*, e1063. [CrossRef]
14. Singh, H.; Thangaraj, P.; Chakrabarti, A. *Acinetobacter baumannii*: A Brief Account of Mechanisms of Multidrug Resistance and Current and Future Therapeutic Management. *J. Clin. Diagn. Res.* **2013**, *7*, 2602–2605. [CrossRef] [PubMed]
15. Nikaido, H.; Takatsuka, Y. Mechanisms of RND multidrug efflux pumps. *Biochim. Biophys. Acta* **2009**, *1794*, 769–781. [CrossRef]
16. Leus, I.V.; Adamiak, J.; Trinh, A.N.; Smith, R.D.; Smith, L.; Richardson, S.; Ernst, R.K.; Zgurskaya, H.I. Inactivation of AdeABC and AdeIJK efflux pumps elicits specific nonoverlapping transcriptional and phenotypic responses in *Acinetobacter baumannii*. *Mol. Microbiol.* **2020**, *114*, 1049–1065. [CrossRef]

17. Damier-Piolle, L.; Magnet, S.; Brémont, S.; Lambert, T.; Courvalin, P. AdeIJK, a Resistance-Nodulation-Cell Division Pump Effluxing Multiple Antibiotics in *Acinetobacter baumannii*. *Antimicrob. Agents Chemother.* **2008**, *52*, 557–562. [CrossRef]
18. Coyne, S.; Guigon, G.; Courvalin, P.; Périchon, B. Screening and Quantification of the Expression of Antibiotic Resistance Genes in *Acinetobacter baumannii* with a Microarray. *Antimicrob. Agents Chemother.* **2010**, *54*, 333–340. [CrossRef]
19. Zaki, M.E.S.; ElKheir, N.A.; Mofreh, M. Molecular Study of Quinolone Resistance Determining Regions of gyrA Gene and parC Genes in Clinical Isolates of *Acintobacter baumannii* Resistant to Fluoroquinolone. *Open Microbiol. J.* **2018**, *12*, 116–122. [CrossRef]
20. Alkasaby, N.M.; Zaki, M.E.S. Molecular Study of *Acinetobacter baumannii* Isolates for Metallo-β-Lactamases and Extended-Spectrum-β-Lactamases Genes in Intensive Care Unit, Mansoura University Hospital, Egypt. *Int. J. Microbiol.* **2017**, *2017*, 3925868. [CrossRef]
21. Al-Agamy, M.H.; Khalaf, N.G.; Tawfick, M.M.; Shibl, A.M.; Kholy, A.E. Molecular characterization of carbapenem-insensitive *Acinetobacter baumannii* in Egypt. *Int. J. Infect. Dis.* **2014**, *22*, 49–54. [CrossRef] [PubMed]
22. Sekyere, J.O.; Amoako, D.G. Carbonyl Cyanide m-Chlorophenylhydrazine (CCCP) Reverses Resistance to Colistin, but Not to Carbapenems and Tigecycline in Multidrug-Resistant Enterobacteriaceae. *Front. Microbiol.* **2017**, *8*, 228. [CrossRef] [PubMed]
23. Lin, L.; Ling, B.-D.; Li, X.-Z. Distribution of the multidrug efflux pump genes, adeABC, adeDE and adeIJK, and class 1 integron genes in multiple-antimicrobial-resistant clinical isolates of *Acinetobacter baumannii*–Acinetobacter calcoaceticus complex. *Int. J. Antimicrob. Agents* **2009**, *33*, 27–32. [CrossRef] [PubMed]
24. Roy, S.; Chatterjee, S.; Bhattacharjee, A.; Chattopadhyay, P.; Saha, B.; Dutta, S.; Basu, S. Overexpression of Efflux Pumps, Mutations in the Pumps' Regulators, Chromosomal Mutations, and AAC(6′)-Ib-cr Are Associated with Fluoroquinolone Resistance in Diverse Sequence Types of Neonatal Septicaemic *Acinetobacter baumannii*: A 7-Year Single Center Study. *Front. Microbiol.* **2021**, *12*, 602724. [CrossRef] [PubMed]
25. He, X.; Lu, F.; Yuan, F.; Jiang, D.; Zhao, P.; Zhu, J.; Cheng, H.; Cao, J.; Lu, G. Biofilm Formation Caused by Clinical *Acinetobacter baumannii* Isolates Is Associated with Overexpression of the AdeFGH Efflux Pump. *Antimicrob. Agents Chemother.* **2015**, *59*, 4817–4825. [CrossRef] [PubMed]
26. Poole, K. Outer Membranes and Efflux: The Path to Multidrug Resistance in Gram—Negative Bacteria. *Curr. Pharm. Biotechnol.* **2002**, *3*, 77–98. [CrossRef] [PubMed]
27. Gallagher, P.; Baker, S. Developing new therapeutic approaches for treating infections caused by multi-drug resistant *Acinetobacter baumannii*: *Acinetobacter baumannii* therapeutics. *J. Infect.* **2020**, *81*, 857–861. [CrossRef]
28. Amin, M.; Navidifar, T.; Shooshtari, F.S.; Rashno, M.; Savari, M.; Jahangirmehr, F.; Arshadi, M. Association Between Biofilm Formation, Structure, and the Expression Levels of Genes Related to biofilm formation and Biofilm-Specific Resistance of *Acinetobacter baumannii* Strains Isolated from Burn Infection in Ahvaz, Iran. *Infect. Drug Resist.* **2019**, *12*, 3867–3881. [CrossRef]
29. Choquet, M.; Lohou, E.; Pair, E.; Sonnet, P.; Mullié, C. Efflux Pump Overexpression Profiling in *Acinetobacter baumannii* and Study of New 1-(1-Naphthylmethyl)-Piperazine Analogs as Potential Efflux Inhibitors. *Antimicrob. Agents Chemother.* **2021**, *65*, e00710-21. [CrossRef]
30. Pérez-Varela, M.; Corral, J.; Aranda, J.; Barbé, J. Functional Characterization of AbaQ, a Novel Efflux Pump Mediating Quinolone Resistance in *Acinetobacter baumannii*. *Antimicrob. Agents Chemother.* **2018**, *62*, e00906-18. [CrossRef]
31. Su, X.-Z.; Chen, J.; Mizushima, T.; Kuroda, T.; Tsuchiya, T. AbeM, an H^+-Coupled *Acinetobacter baumannii* Multidrug Efflux Pump Belonging to the MATE Family of Transporters. *Antimicrob. Agents Chemother.* **2005**, *49*, 4362–4364. [CrossRef]
32. Leus, I.V.; Weeks, J.W.; Bonifay, V.; Smith, L.; Richardson, S.; Zgurskaya, H.I. Substrate Specificities and Efflux Efficiencies of RND Efflux Pumps of *Acinetobacter baumannii*. *J. Bacteriol.* **2018**, *200*, e00049-18. [CrossRef] [PubMed]
33. Richmond, G.E.; Evans, L.P.; Anderson, M.J.; Wand, M.E.; Bonney, L.C.; Ivens, A.; Chua, K.L.; Webber, M.A.; Sutton, J.M.; Peterson, M.L.; et al. The *Acinetobacter baumannii* Two-Component System AdeRS Regulates Genes Required for Multidrug Efflux, Biofilm Formation, and Virulence in a Strain-Specific Manner. *Mbio* **2016**, *7*, e00430-16. [CrossRef] [PubMed]
34. Yoon, E.-J.; Courvalin, P.; Grillot-Courvalin, C. RND-Type Efflux Pumps in Multidrug-Resistant Clinical Isolates of *Acinetobacter baumannii*: Major Role for AdeABC Overexpression and AdeRS Mutations. *Antimicrob. Agents Chemother.* **2013**, *57*, 2989–2995. [CrossRef] [PubMed]
35. Yoon, E.-J.; Balloy, V.; Fiette, L.; Chignard, M.; Courvalin, P.; Grillot-Courvalin, C. Contribution of the Ade Resistance-Nodulation-Cell Division-Type Efflux Pumps to Fitness and Pathogenesis of *Acinetobacter baumannii*. *Mbio* **2016**, *7*, e00697-16. [CrossRef] [PubMed]
36. Chen, H.; Cao, J.; Zhou, C.; Liu, H.; Zhang, X.; Zhou, T. Biofilm Formation Restrained by Subinhibitory Concentrations of Tigecyclin in *Acinetobacter baumannii* Is Associated with Downregulation of Efflux Pumps. *Chemotherapy* **2017**, *62*, 128–133. [CrossRef]
37. Alav, I.; Sutton, J.M.; Rahman, K.M. Role of bacterial efflux pumps in biofilm formation. *J. Antimicrob. Chemother.* **2018**, *73*, 2003–2020. [CrossRef]
38. Hershberg, R. Antibiotic-Independent Adaptive Effects of Antibiotic Resistance Mutations. *Trends Genet.* **2017**, *33*, 521–528. [CrossRef]
39. Jung, H.-W.; Kim, K.; Islam, M.M.; Lee, J.C.; Shin, M. Role of ppGpp-regulated efflux genes in *Acinetobacter baumannii*. *J. Antimicrob. Chemother.* **2020**, *75*, 1130–1134. [CrossRef]
40. Lin, M.-F.; Lin, Y.-Y.; Yeh, H.-W.; Lan, C.-Y. Role of the BaeSR two-component system in the regulation of *Acinetobacter baumannii* adeAB genes and its correlation with tigecycline susceptibility. *BMC Microbiol.* **2014**, *14*, 119. [CrossRef]

41. Henry, R.; Crane, B.; Powell, D.; Lucas, D.D.; Li, Z.; Aranda, J.; Harrison, P.; Nation, R.L.; Adler, B.; Harper, M.; et al. The transcriptomic response of *Acinetobacter baumannii* to colistin and doripenem alone and in combination in an in vitro pharmacokinetics/pharmacodynamics model. *J. Antimicrob. Chemother.* **2015**, *70*, 1303–1313. [CrossRef] [PubMed]
42. Fernando, D.; Zhanel, G.; Kumar, A. Antibiotic Resistance and Expression Of Resistance-Nodulation-Division Pump- and Outer Membrane Porin-Encoding Genes inAcinetobacterSpecies Isolated from Canadian Hospitals. *Can. J. Infect. Dis. Med. Microbiol.* **2013**, *24*, 17–21. [CrossRef] [PubMed]
43. Amer, M.; Wasfi, R.; Attia, A.; Ramadan, M. Indole Derivatives Obtained from Egyptian *Enterobacter* sp. Soil Isolates Exhibit Antivirulence Activities against Uropathogenic *Proteus mirabilis*. *Antibiotics* **2021**, *10*, 363. [CrossRef] [PubMed]
44. Yaikhan, T.; Chuerboon, M.; Tippayatham, N.; Atimuttikul, N.; Nuidate, T.; Yingkajorn, M.; Tun, A.W.; Buncherd, H.; Tansila, N. Indole and Derivatives Modulate Biofilm Formation and Antibiotic Tolerance of Klebsiella pneumoniae. *Indian J. Microbiol.* **2019**, *59*, 460–467. [CrossRef]
45. Sanchez-Carbonel, A.; Mondragón, B.; López-Chegne, N.; Peña-Tuesta, I.; Huayan-Dávila, G.; Blitchtein, D.; Carrillo-Ng, H.; Silva-Caso, W.; Aguilar-Luis, M.A.; del Valle-Mendoza, J. The effect of the efflux pump inhibitor Carbonyl Cyanide m-Chlorophenylhydrazone (CCCP) on the susceptibility to imipenem and cefepime in clinical strains of *Acinetobacter baumannii*. *PLoS ONE* **2021**, *16*, e0259915. [CrossRef] [PubMed]
46. Ikonomidis, A.; Tsakris, A.; Kanellopoulou, M.; Maniatis, A.; Pournaras, S. Effect of the proton motive force inhibitor carbonyl cyanide-*m*-chlorophenylhydrazone (CCCP) on *Pseudomonas aeruginosa* biofilm development. *Lett. Appl. Microbiol.* **2008**, *47*, 298–302. [CrossRef] [PubMed]
47. Zhang, Y.; Li, Z.; He, X.; Ding, F.; Wu, W.; Luo, Y.; Fan, B.; Cao, H. Overproduction of efflux pumps caused reduced susceptibility to carbapenem under consecutive imipenem-selected stress in *Acinetobacter baumannii*. *Infect. Drug Resist.* **2018**, *11*, 457–467. [CrossRef]

antibiotics

MDPI

Article

Complex Infection-Control Measures with Disinfectant Switch Help the Successful Early Control of Carbapenem-Resistant *Acinetobacter baumannii* Outbreak in Intensive Care Unit

Jozsef Kelemen [1], Marton Sztermen [1], Eva Krisztina Dakos [1], Jozsef Budai [1], Jozsef Katona [1], Zsuzsanna Szekeressy [2], Laszlo Sipos [1], Zoltan Papp [1], Balazs Stercz [3,4], Zsuzsanna A. Dunai [4], Bela Kocsis [3], Janos Juhasz [3,5], Fruzsina Michelisz [1], Zsuzsanna Daku [1], Judit Domokos [3,4], Dora Szabo [1,3,4,*,†] and Lorand Eross [1,†]

1 Department of Neurosurgery and Neurointervention, Semmelweis University, 1083 Budapest, Hungary
2 Gyula Nyírő National Institute of Psychiatry and Addiction, 1135 Budapest, Hungary
3 Institute of Medical Microbiology, Semmelweis University, 1089 Budapest, Hungary
4 HUN-REN-SU Human Microbiota Research Group, 1052 Budapest, Hungary
5 Faculty of Information Technology and Bionics, Péter Pázmány Catholic University, 1083 Budapest, Hungary
* Correspondence: szabo.dora@semmelweis.hu
† These authors contributed equally to this work.

Abstract: A carbapenem-resistant *Acinetobacter baumannii* (CRAB) outbreak in an intensive care unit (ICU) was contained by an improved infection-control measure that included a disinfectant policy. In our retrospective cohort study, we describe the epidemiological investigations and infection-control measures during this outbreak. Descriptive analysis was used to summarize patient demographics, neurological diseases, surgical treatment, underlying diseases, infection, and outcomes. In December 2023, two CARB-positive patients were observed in the ICU, and four more patients became CRAB-positive in January. During this outbreak, there was an overlap of hospitalization periods among the CRAB-positive patients, and CRAB was isolated from the environment; the isolated CRAB strain was identical. Infection-control measures, including hand hygiene, contact precautions and isolation, surveillance, decolonization, environmental cleaning, and disinfection, were reviewed and modified. The aim of this study was to examine the molecular background of the effectiveness of the disinfectant shift used during successful outbreak control. Experiments were carried out to study the phenotypic sensitivity and genetic background of different disinfectant agents. A thorough analysis of the detected CRAB strain included whole-genome sequencing (WGS), investigation of the *qacE* and *qacE∆1* genes' relative expression by qPCR after exposure to different disinfectant solutions, as well as an analysis of biofilm formation. WGS analysis of the CRAB strain identified that an ST2 high-risk clone was responsible for the outbreak, which produced OXA-83 and ADC-30 beta-lactamases; in addition, *qacE* and *qacE∆1* genes were also detected, which confer resistance to disinfectants containing quaternary ammonium compounds (QACs). A qPCR analysis demonstrated that after exposure to different disinfectants, the gene expression levels of *qacE* and *qacE∆1* increased and correlated with concentrations of QACs of disinfectants. During the outbreak, the standard-of-care QAC-based disinfectant was changed to a mainly alcohol-based agent in the ICU, which contributed to the successful control of this outbreak, and no additional patients were identified with CRAB. We conclude that continuous surveillance and hand hygiene training combined with fast identification and reaction to new cases, as well as an in-depth analysis of multidrug-resistant outbreak strains and investigation of their disinfectant tolerance/resistance during an outbreak, are essential to effectively control the spread of nosocomial pathogens. The smart policy of disinfectant agent selection played a crucial role in controlling the outbreak and ensuring patient safety in the ICU.

Keywords: carbapenem-resistant *Acinetobacter baumannii*; healthcare-associated infection; HAI; antimicrobial resistance; disinfectant resistance; infection control; infection prevention; multidrug-resistant organism; outbreak; intensive care unit

1. Introduction

Outbreaks induced by multidrug-resistant *Acinetobacter baumannii* have been reported in recent years with an increasing frequency, representing an alarming threat in hospital settings worldwide [1,2]. *A. baumannii* is a major nosocomial pathogen, as it causes various systemic and life-threatening infections (e.g., wound infections, bloodstream infections, urinary tract infections, ventilator-associated pneumonia, etc.) among hospitalized patients. *A. baumannii* belongs to the ESKAPE group of infections, which are difficult to treat and are associated with high mortality rates, as only a limited number of effective antibacterial agents are available to treat such infections [3–10]. The World Health Organization has published a priority list where carbapenem-resistant *A. baumannii* (CRAB) is in the critical group; therefore, research, development, and discovery of new antibacterial agents are needed in order to treat these infections [11].

The genome of *A. baumannii* has high variability, as it can acquire and accumulate different antibiotic-resistance genes; thus, multidrug-resistant strains can develop. Nowadays, the major concern is about CRAB, that additional high-risk *A. baumannii* clones could evolve as they exhibit multidrug resistance and have the ability to survive and persist over a longer term in the environment (e.g., hospital environments). The most frequently reported high-risk *A. baumannii* clones belong to the following sequence types (STs): ST1, ST2, and ST3. Having said that, the currently predominant high-risk clone worldwide is ST2 [12].

Carbapenem resistance in *A. baumannii* is mainly explained by the production of OXA-type beta-lactamases (e.g., OXA-23-like, OXA-24/40-like, OXA-51-like, and OXA-58-like), but metallo-beta-lactamases (e.g., NDM, IMP, and VIM) can also be acquired by horizontal gene transfer, and among them, the NDM type is the most frequently reported in *A. baumannii* [13–15]. Additionally, resistance to fluoroquinolones, aminoglycosides, and colistin has been reported worldwide with an increasing frequency [16–18]. Furthermore, resistance to disinfectants containing quaternary ammonium compounds (QACs) and antiseptics has also been scrutinized [19–21]. This is especially true since QACs are commonly used disinfectants that include a high number of different agents such as benzalkonium chloride, didecyl dimethylammonium chloride, and alkyl trimethyl ammonium chloride [21].

Biofilms are resilient structures that are widely found in various natural habitats. They are also present in healthcare settings, where they are associated with healthcare-associated infections (HAIs). Healthcare-setting-associated biofilms are generally less susceptible to antimicrobials and disinfectants that are primarily used to inhibit the initial formation than others applied in less challenging environments. Eliminating established biofilms often requires additional, more aggressive disinfectant strategies on top of regular cleaning agents [22]. Surface biofilms with multi-resistant bacteria are common sources of environmental pathogens and therefore are associated with HAIs. Additional factors also influence the effectiveness of disinfectants, cumulatively decreasing the efficiency of infection prevention and control [22].

Choosing the appropriate disinfectant is crucial for effective infection control. Disinfectants recommended for patient-care items include QACs, alcohol, glutaraldehyde, hydrogen peroxide, peracetic acid, sodium hypochlorite, iodophors, and phenolics. Further important parameters need to be specified to ensure effective and safe infection control and prevent HAIs at sufficient disinfectant concentrations and exposure times. In addition, understanding their mechanisms of action and proper administration are needed to reduce the infection risk of vulnerable patients.

It is important to highlight the importance of touchless, i.e., contact-free techniques or techniques with minimal direct contact in cleaning and disinfection. These are useful practices to reduce cross-contamination and improve hygiene and safety in healthcare facilities and other cleanliness-critical environments. Given the increased tolerance and resistance to disinfectants, non-contact disinfection techniques are coming to the forefront. Furthermore, the successful use of innovative techniques—for example, violet-blue LED lamps—is frequently reported [23,24].

The purpose of the current study was to determine the role of disinfectant resistance genes in the spread of a CRAB strain during the outbreak, given that CRAB outbreaks are usually prolonged, persisting for months or years in hospitals [25,26]. In this report, we present successful and rapid infection-control measures focusing on disinfectant resistance in the control of a CRAB outbreak in a neurosurgery intensive care unit and the subsequent strengthening of infection-control interventions.

2. Results

2.1. Description of the Outbreak

Between 28 December 2023 and 22 February 2024, CRAB strains were isolated from six patients in the neurosurgical ICU at the Department of Neurosurgery and Neurointervention, Semmelweis University, Budapest, Hungary (Figure 1). The hospitalization periods of the CRAB-positive patients overlapped. The incidence of CRAB cases rapidly increased in December 2023 and moderately decreased in the first two weeks of January 2024. A rapid decrease in incidence was observed in the third week of January (Figure 1). Relevant background information on the patients treated in the neurosurgical ICU during this period is summarized in Table 1, and the characteristics of the six CRAB-positive patients are described in Table 2. Among the six patients infected during the outbreak, two patients were colonized, and four were infected with the newly appeared CRAB strain in the neurosurgical ICU. Five patients underwent craniotomy surgery, and one had a neurointerventional procedure. Clinical specimen sources of CRAB were trachea, blood, urine, liquor, wound, and anorectal swab. Of the six patients involved in this outbreak, three recovered from the CRAB infection; however, one patient died of a secondary bloodstream infection after a primary wound infection. Given that the patient with CRAB-positive wound infection was a contact person, we assumed that the CRAB wound infection was acquired in the ICU.

Table 1. Characteristics of all the patients treated in the neurosurgical ICU during the study period.

Variables	All Patients (n = 291)
Age (years)	55.4 (38–85)
Gender, male	135 (46%)
Neurosurgical disease	
CNS tumor	175 (60%)
Subarachnoid hemorrhage	62 (21%)
Other hemorrhages	22 (46%)
Stroke	32 (7%)
Surgical treatment	
Neurosurgical operation	207 (71%)
Endovascular treatment	84 (29%)
Underlying diseases	
Hypertension	42 (14%)
Diabetes mellitus	49 (16%)
Others	64 (22%)
Infection	
Urinary tract infection	34 (11.6%)
Pneumonia	19 (7%)
Wound infection	10 (3.4%)
Bloodstream infection	7 (2%)
Meningitis	3 (1%)
Death	12 (4%)

CNS: central nervus system.

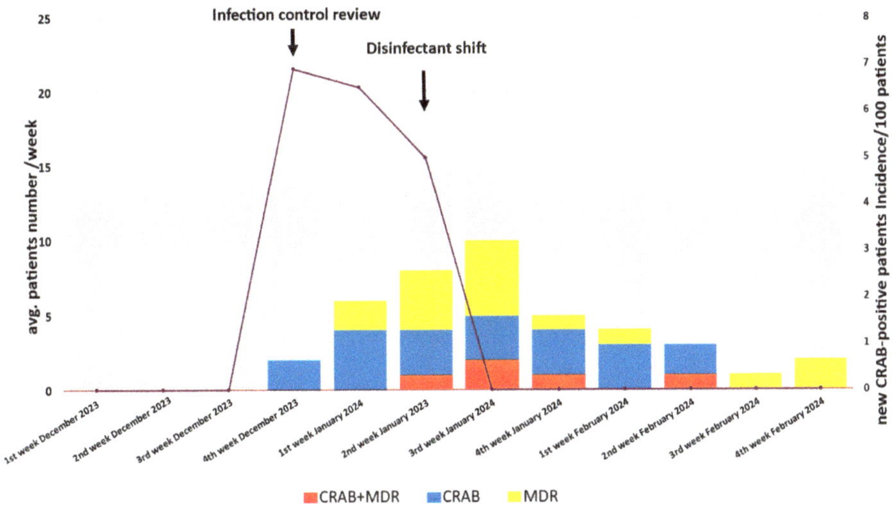

Figure 1. Timeline of the CRAB outbreak. The gray boxes indicate the number of patients treated each week in the ICU. The blue boxes represent patients with only CRAB.

Table 2. Characteristics of CRAB-positive patients.

Patient	Age (Years)	Gender	Underlying Diseases	Surgical Procedure	Multiresistant Bacteria	CRAB-Positive Specimens	Infection/ Colonization	Antibiotic Treatment	Duration of Hospital Stay (Days)	Outcome
P1	37	male	intracranial hemorrhage, hypertension	craniotomy	CRAB, MRSA	trachea	colonization	imipenem /cilastatin	29	discharged
P2	63	male	subarachnoid hemorrhage, hypertension	craniotomy	CRAB, VRE	urine, liquor, anorectal	infection: UTI, meningitis	colomycin, imipenem /cilastatin	58	discharged
P3	48	female	stroke, hypertension	intervention	CRAB	bloodculture	infection: BSI	colomycin	12	discharged
P4	67	female	AVM angiomatis, hypertension, diabetes mellitus, COPD	craniotomy	CRAB	trachea	colonization	imipenem /cilastatin, vancomycin, meropenem	6	discharged
P5	76	female	cerebral metastasis, hypertension, hypothyreosis, breast cancer	craniotomy	CRAB, ESBL-KP, VRE	wound, nasal, bloodculture	infection: SSI, BSI	colomycin, meropenem, vancomycin, linezolid, gentamycin	36	exit
P6	53	male	subarachnoid hemorrhage, hypertension, diabetes mellitus	craniotomy	CRAB, ESBL-KP, VRE	trachea, bloodculture, anorectal	infection: VAP, BSI	colomycin, ceftriaxone, cefatizidime, imipenem /cilastatin, meropenem, levofloxacin	38	discharged

CRAB: carbapenem-resistant *Acinetobacter baumannii*, MRSA: methicillin-resistant Staphylococcus aureus, VRE: vancomycin-resistant Enterococcus, ESBL-KP: extended-spectrum beta-lactamase producing K. pneumoniae. AVM: arteriovenous malformation, COPD: chronic obstructive pulmonary disease; UTI: urinary tract infection, BSI: bloodstream infection, VAP: ventilator-associated pneumonia.

2.2. Environmental Investigations

There were no pathogen bacteria isolated from the surgery rooms and from any related surfaces, medical devices, or sinks. In contrast, at the ICU, different multidrug-resistant bacteria, including vancomycin-resistant Enterococcus (VRE) and extended-spectrum beta-lactamase (ESBL)-producing bacteria were detected on the different surfaces. VRE was isolated from the patient room surfaces, sink, sink drain, and excreta. The outbreak CRAB strain was detected in the excreta discharge, even after a few days of taking the samples. However, no CRAB was detected in follow-up tests conducted until June 2024.

2.3. Infection-Control Measures during the Outbreak

Upon the onset of this CRAB outbreak, existing infection-control interventions were reviewed, and a set of containment measures were instituted. Table 3 summarizes the infection-control measures used before, during, and after the CRAB outbreak.

Table 3. Summary of the infection-control strategy for CRAB outbreak.

	Before CRAB Outbreak	During CRAB Outbreak	After CRAB Outbreak
Enforcement of hand hygiene			
Regular education programs on hand hygiene	yearly	weekly, including circulating staff	every six months
Audit of hand hygiene compliance	every six month	daily	weekly
Universal contact precautions			
earing gloves, coat, apron, and shoe protector in the patient zone	in case of MDR-positive patients	in case of MDR-positive patients and their contacts	in case of MDR-positive patients and their contacts
Wearing of personal protective equipment by HCWs	yes	yes	yes
Access restrictions	yes	yes	yes
Review of invasive procedures	yes	yes	yes
Cohort isolation			
cohort isolation of colonized patient	no	yes	yes
Healthcare worker cohorting (HCW), patient:HCW ratio	3:1	2:1	3:1
Surveillance cultures			
Passive surveillance: antibiogram monitoring	yes	yes	yes
Active surveillance: screening cultures	no	initial (in 24 h) and in every 5 days	initial (in 24 h) and weekly
Environmental surveillance cultures	occasionally	once a week	every three months
Patient decolonization			
Throat decolonizitation with chlorhexidine digluconate	no	yes	yes
Environmental cleaning and disinfection			
Extensive management using checklists under the supervision of the unit manager	no	yes	yes
Patient zone	twice a day	three times a day, and necessery	three times a day, and necessery
Environment except the patient zone	twice a day	three times a day	three times a day
Cleaning personnel	two persons for the unit	three cohorting cleaning persons for the unit	three cohorting cleaning persons for the unit
Improved waste management	yes	yes	yes
Touchless environmental disinfection systems (e.g., vaporized hydrogen peroxide)	yes	yes	yes
Developing a purposeful disinfectant policy			
Disinfection stewardship	no	**Disinfection switch**	yes
Antibiotic stewardship	yes	yes	yes
Genomic surveillance			
Whole-genome sequencing	no	yes	no
Monitoring and feedback			
Discussions on the performance of the infection-control strategy	monthly	daily	weekly

During the CRAB outbreak, several measures were implemented to control the spread of the multidrug-resistant pathogen (Table 3). Hand hygiene education programs were held weekly, with daily monitoring of compliance. CRAB patients and positive contacts were isolated after diagnosis and occasionally housed in single rooms at the ICU. The patient:healthcare worker ratio was improved to 2:1 during the outbreak. Access by healthcare workers to the ICU was limited, and cohorting of patients and healthcare workers was implemented.

Throat decolonization with chlorhexidine digluconate was introduced. Environmental surveillance tests were conducted weekly in the ICU and surgery rooms.

Cleaning and disinfection procedures were enhanced with a checklist and increased frequency. The fourth-generation QAC disinfectant—didecyl dimethyl-ammonium chloride—has been constantly used in the surgery room. However, in the ICU, a first-generation QAC disinfectant—Alkyl (C12–16) dimethyl benzyl ammonium chloride (ADBAC [C12–18]), also known as MBF—was originally used. Patient zones were cleaned three times a day using both touch and touchless techniques.

In the ICU, a new disinfectant, named IP—predominantly containing 2-phenoxy-ethanol (10–20%) and benzalkonium chloride as active ingredients at 5–10% as a first-generation QAC component—replaced the previously used MBF disinfectant after a CRAB-positive patient and environmental sample were identified.

The introduction of the new disinfectant, IP, resulted in no new cases of CRAB. This intervention, together with environmental cleaning, disinfection, and surveillance, helped prevent the spread of CRAB during the outbreak.

2.4. *Characteristics of CRAB Strains of This Outbreak*

2.4.1. Biofilm-Forming Capacity

According to the breakpoints observed using the microtiter-based method, the outbreak CRAB strains were found to be weak biofilm producer isolates.

2.4.2. Antibiotic Susceptibility

Multidrug-resistant *A. baumannii* strains were isolated from six patients and from the environment during the outbreak. Antibiotic susceptibility testing revealed that all *A. baumannii* strains, both from clinical and environmental samples, exhibited identical resistance patterns. They were resistant to carbapenems, aminoglycosides, and fluoroquinolones but were susceptible to colistin with a 1 μg/mL minimum inhibitory concentration (MIC) value.

2.4.3. Disinfectant Susceptibility of the Planktonic and Biofilm-Forming Phase of CRAB

The MIC and minimum bactericide concentration (MBC) were determined to be benzalkonium chlorate, 2-phenoxyethanol, and chlorhexidine digluconate for the planktonic phase of the outbreak CRAB strain. MIC values for both *A. baumannii* strains (the outbreak CRAB) and control *A. baumannii* BAA-1805 (purchased from ATCC and isolated from a Canadian soldier injured in Afghanistan) [27] were high, 16–32 μg/mL for benzalkonium chloride and chlorhexidine digluconate, and as expected, the MBC values were even higher, 32–64 μg/mL as well. However, the MIC values of 2-phenoxyethanol were significantly lower, 0.25 μg/mL for both the outbreak CRAB and for the control *A. baumannii* strains. For all investigated disinfectant agents, the MBC/MIC ratio was 1 or 2, indicating a bactericidal effect (Table 4).

Since the outbreak CRAB strain was found to be a weak biofilm producer, the minimum biofilm inhibitory concentration (MBIC) was determined. Benzalkonium chloride, 2-phenoxyethanol, and chlorhexidine digluconate inhibited biofilm formation of the outbreak CRAB strain at low concentrations. The phenomenon observed for MIC values was also observed for MBIC. The lowest MBIC value was measured for 2-phenoxy-ethanol, as 2phenoxy-ethanol inhibited biofilm formation to the greatest extent (Table 4).

Table 4. The minimum inhibitory concentration (MIC), the minimum bactericidal concentration (MBC), and the minimum biofilm inhibitory concentration (MBIC) of the outbreak CRAB strain (µg/mL).

Disinfectants	*Acinetobacter baumannii* Outbreak CRAB				*Acinetobacter baumannii* BAA-1805			
	MIC	MBC	MBC/MIC	MBIC	MIC	MBC	MBC/MIC	MBIC
Benzalkonium chloride	32	32	1	0.5	16	32	2	<0.06
2-phenoxyethanol	0.5	0.25	2	0.0078	0.25	0.25	1	0.0078
Chlorhexidine digluconate	16	32	2	0.25	32	64	2	8

2.4.4. Killing Curve Assay with the Two Different Surface Disinfectants—MBF and IP—Used during the Outbreak

An in vitro killing curve assay was performed to determine the growth capacity of the planktonic CRAB strain. The growth of the CRAB strain was investigated in the presence of two types of disinfectants. MBF, which contained a higher percentage (10–20%) of first-generation QAC, was used as surface disinfectant before the outbreak. During the outbreak, it was switched to IP, the active ingredient of which was 2-phenoxy-ethanol (10–20%), but it also contained 5–10% of first-generation QAC compounds. Upon exposure to sub-lethal concentrations of both MBF and IP surface disinfectant agents, the CRAB outbreak strain was able to grow at 0.039% and at 0.078% (Figure 2).

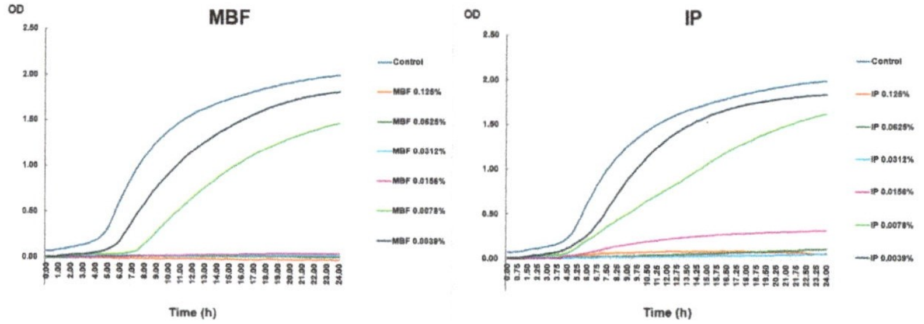

Figure 2. Killing curve assay.

The growth curve of the planktonic CRAB strain is shown under different concentrations of surface disinfectants over 24 h. Bacterial growth was detected by measuring the OD. The figure's left panel shows the rate of growth of different concentrations of MBF, and the right panel shows the increase in the values of IP.

2.4.5. Genomic Analysis

In order to reveal the additional properties of the CRAB outbreak strain, whole-genome analysis was performed, and different antibiotic resistance genes were detected (Figure 3). OXA-83 and ADC-30 beta-lactamases, *aph(3″)-I*, *aph(3″)-Ia*, *aph(6)-Id*, and *aac(6″)-Ia* aminoglycoside resistance determinants, were identified. Furthermore, several efflux pump genes were observed to induce resistance against many antimicrobial agents. Additionally, *pgaA*, *pgaB*, *pgaC*, *bfmR*, *bfmS*, and *ompA* genes playing a role in biofilm formation were recognized (Table 5). Interestingly, acquired resistance genes, *qacE* and its variant *qacEΔ1*, were detected in the genome of CRAB. These genes code resistance to disinfectants that contain QACs.

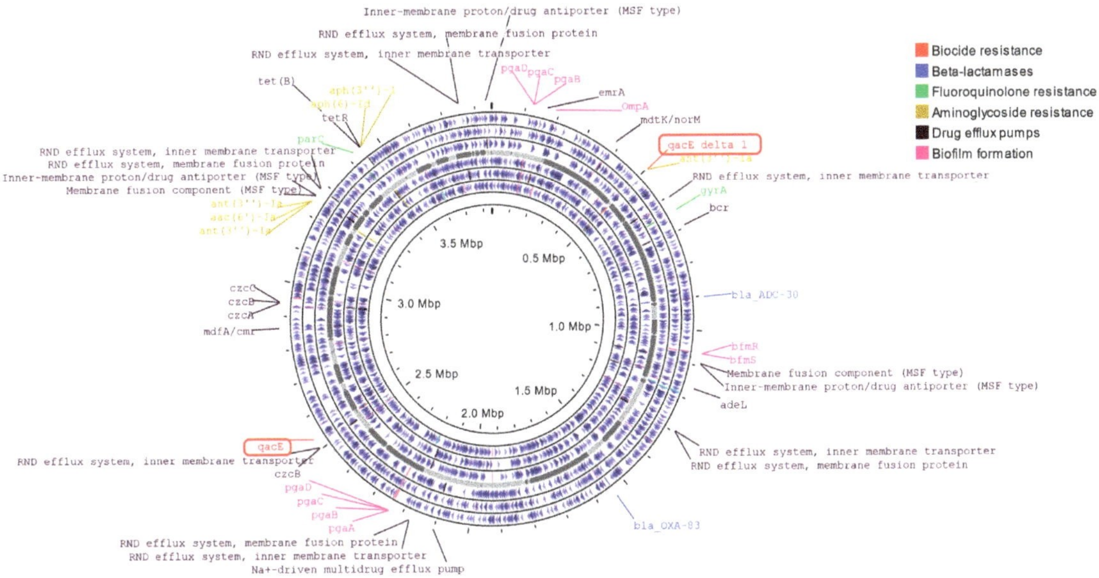

Figure 3. Whole-genome sequence analysis of *A. baumannii* strain of this outbreak.

Table 5. Characteristics of genome sequence of *A. baumannii* (JBEVZX010000001).

Multilocus sequence typing (Pasteur schema)	ST2
Beta-lactamases	*bla*$_{\text{OXA-83}}$ (OXA-51-like)
	bla$_{\text{ADC-30}}$
Fluoroquinolone resistance determinants	*gyrA*: Ser81Leu *parC*: Ser84Leu
Aminoglycoside resistance genes	*ant(3″)-II* (aminoglycoside nucleotidyltransferase) *aph(6)-Id* (aminoglycoside O-phosphotransferase) *aph(3″)-Ib* (aminoglycoside O-phosphotransferase) *aac(6′)-Ia* (aminoglycoside 6′-N-acetyltransferase)
Efflux pumps	*adeABC; adeFGH; adeIJK; adeRS; adeL; adeN*
Disinfectant, antiseptic resistance determinants	*qacE; qacEΔ1*
Biofilm formation	*pgaABCD; csuA/B, csuABCDE; bfmR; bfmS; ompA*

2.4.6. Real-Time PCR Assay to Determine the Relative Expression Rate of *qacE* and *qacEΔ1*

Since at sub-lethal concentrations of 0.039% and 0.078% in the presence of both MBF and IP surface disinfectant agents, the CRAB outbreak strain was still able to grow, these two concentrations were used during the qPCR assay. In order to determine how the *qacE* and *qacEΔ1* genes allow the outbreak CRAB bacteria to adapt and develop tolerance to QACs at sub-lethal concentrations, real-time PCR was used to quantify the *qacE* and *qacEΔ1* relative expression rate. As a housekeeping gene, 16S was used. A sub-lethal 0.0078% concentration of both MBF and IP surface disinfectant agents hardly changed the relative expression rate of either *qacE* or *qacEΔ1*. However, at a sub-lethal 0.0039% concentration, MBF surface disinfectant agents significantly increased the relative expression of both *qacE* and *qacEΔ1*. The relative expression of the *qacE* gene increased six times, and that of the *qacEΔ1* gene increased >9 times (Figure 4). The simultaneous presence of the two genes, *qacE* and *qacEΔ1*, provides an additive effect.

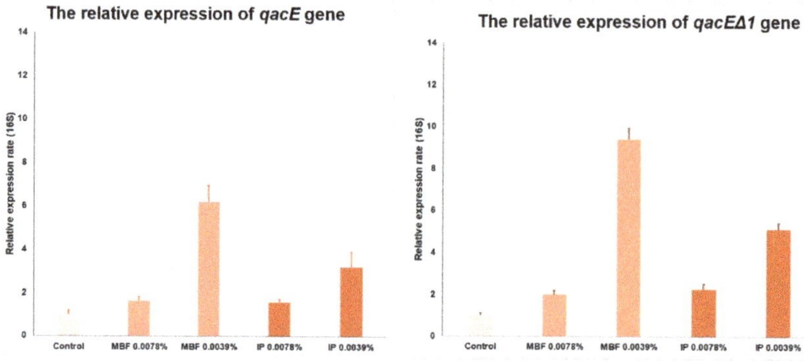

Figure 4. The real-time PCR assay.

The circos plot shows the GC content skew (inner circle with histograms). The SPAdes assembled contigs are plotted in gray (middle circle). The positions of all predicted genes are shown separately by reading frames (six circles with blue marks).

The relative expression of *qacE* (on the left) and *qacEΔ1* (on the right) genes in different concentrations, 0.0078% and 0.0039%, of the two surface disinfectants—MBF and IP—is shown.

Based on the relative expression, apparently, the expression of *qacE* and *qacEΔ1* genes are concentration-dependent and inducible. CRAB showed an increase at both 0.0078% and 0.0039% MBF and IP concentrations, but the expression of the *qacE* and *qacEΔ1* efflux pumps increased only at a 0.0039% concentration. The dependence of the relative expression on the concentration and active substance content is further confirmed by the fact that the difference between the increase in MBF and IP expression is proportional to the QAC content of the two disinfectants. MBF contains 10–20% and IP 5–10% QAC, and the *gacE* expression is increased by 6 times for MBF and by 3.2 times for IP. These values regarding *qacEΔ1* expression are increased 9.4-fold for MBF and 5.1-fold for IP.

3. Discussion

HAIs are commonplace but preventable events, affecting patients in healthcare settings worldwide. Enhancing the routine cleaning and disinfection procedures of hospital environments has been proven to lower the risk of HAIs [28]. The prevention and control of HAIs represent a major global public health challenge, raising concerns among healthcare workers, patients, and the public. This issue has become even more critical with the rise of multidrug-resistant pathogens.

The global incidence of antibiotic resistance in *A. baumannii* has increased steadily in the past 10 years, and it is associated with a higher burden of attributed mortality [29,30]. CRAB is a major concern in healthcare settings due to its high antibiotic resistance and persistence and due to its role in several outbreaks worldwide [30–33]. In order to control outbreaks caused by CRAB, several infection-control measures are applied. A multimodal infection prevention control approach appears to be superior, provided it is implemented as a 'bundle' of interventions [34–36].

The purpose of our present study was to determine the factors that led to the rapid and successful eradication of the CRAB outbreak that occurred at the ICU. We clarified the role of the disinfectant shift during outbreak control and investigated the presence and role of genes responsible for disinfectant resistance.

In our study, a multidrug-resistant *A. baumannii* ST2 high-risk clone induced an outbreak in a neurosurgery ICU, and during this outbreak, six patients were involved. We implemented different infection prevention control (IPC) measures, including hand hygiene, contact precautions, staff education, additional active screening, cohorting staff and patients, environmental cleaning and disinfection, monitoring of environmental cleaning, antimicrobial stewardship/monitoring of antibiotic consumption, active respiration, perianal screening, and environmental cultures. The effectiveness of the combination of these techniques has already been described by several studies [37–43]. Nasopharyngeal decolonization with chlorhexidine digluconate introduced during the outbreak in our ICU is another well-documented infection-control measure [44].

Contact precaution is generally applied during CRAB outbreaks; however, staff or nursing cohorting, as we implemented, has rarely been applied [43,45,46]. During IC measures, hand hygiene practices were implemented, and alcohol hand rubs were used; this practice, as a basic practice, is also commonly described in *A. baumannii* outbreaks [47]. Increased hand hygiene control and monitoring also contributed to our outbreak control, as the implementation of hand hygiene best practices was described by other studies as well [41,42,45].

CRAB isolates can survive in the hospital environment for a long time because of biofilm production, so they could be detected from colonized surfaces and environmental samples [47–49]. However, environmental sampling is not widely used and has only been performed in a few reports [43,46,49–51]. In our study, we could successfully culture bacteria from the sink, and we believe that its elimination could play a role in outbreak control.

During the outbreak, passive microbiological surveillance in the ICU was changed to active surveillance, with rectal screening, urine screening, and respiratory–pharynx, trachea–screening tests for all patients upon admission and every 5 days afterward. The high sensitivity of combined screening methods, rectal and pharyngeal screening [41], and rectal and skin screening is well documented [45].

Most infection-control activities are carried out during the control of outbreaks; however, very few studies detail cleaning and disinfection techniques [48]. Opinions differ on the effectiveness of touch and non-touch cleaning techniques. Controlling the efficiency of cleaning and disinfection is questionable and depends on the institution. In a recent review, it was concluded that the choice of disinfectant appears less critical than ensuring thorough cleaning and the complete removal of biofilm [52]. Our data, however, suggest that the disinfectant type used had a strong effect on the outcome of the outbreak. Due to the resistance to disinfectants, properly selected disinfectants have an important role in successful outbreak control. Because of the presence of *qacE* and *qacEΔ1* efflux pump genes and due to their inducibility, we recommend the use of disinfectant agents containing several different active ingredients as a main component to prevent the development of disinfectant resistance.

The correct choice of disinfectant policy, supplemented with additional infection-control measures, contributes significantly to the successful control of CRAB outbreaks.

We managed to control the CRAB outbreak after the introduction of a new, mainly alcohol-based disinfectant IP, which replaced the conventional QAC-based agent MBF. This measure resulted in the containment of the outbreak, and no further infection or dissemination was detected. These results pinpoint the importance of applying disinfectant agents with not only one but multiple active ingredients to render the emergence of resistance.

Unfortunately, it is not possible to routinely characterize the susceptibility to disinfectants of the outbreak strains in hospitals, and there are no breakpoints agreed upon based on international consensus for the standardized breakpoints for the determination of tolerance and resistance. In the future, biocide tolerance and resistance will be recognized sooner by fast detection either by PCR targeting to *qacE* and *qacEΔ1* or by whole-genome sequencing based on the available resources of the laboratory. As a result, appropriate infection-control measures, e.g., changing disinfectants, can take place as soon as possible and can play a significant role in controlling the outbreak.

A recent study analyzed antiseptic MIC values for *A. baumannii* clinical isolates and found that the presence of *qacE*, *qacEΔ1*, and *aceI* genes influence the MIC values of benzalkonium chloride and chlorhexidine digluconate [19]. This observation was confirmed by our study showing that the MIC values were higher for chlorhexidine digluconate and benzalkonium chloride affected by QacE and QacEΔ1 efflux pumps in contrast to low MIC values of 2-phenoxy-ethanol, which is not affected by the efflux pump.

The disinfectant resistance of *A. baumannii* has been reported earlier, and it has been concluded that *qacE* and *qacE*Δ*1* genes in *A. baumannii* confer resistance to QACs [53]. The two genes, *qacE* and *qacE*Δ*1*, encode transmembrane proteins, and these are responsible for adaptation and tolerance to QACs in *A. baumannii*. Additionally, resistance to other types of biocides can be enhanced by them, enabling bacterial pathogens to survive and persist in a hospital environment [20]. It is important that both *qacE* and *qacE*Δ*1* genes are usually located on integrons and on plasmids that can be transferred by horizontal gene transfer between different Gram-negative bacteria (e.g., in *A. baumannii*) [53–55].

As far as we understand, this is the first report on the genes responsible for disinfectant resistance. More precisely, *qacE* and *qacE*Δ*1* were detected during the whole-genome analysis of the outbreak CRAB strain. The *qacE* and qacEΔ1 genes are not only responsible for the observed reduced sensitivity to disinfectants, but QAC disinfectants are able to induce the expression of efflux pumps at a tolerable disinfectant concentration, for which expression induction is concentration-dependent. Understanding this phenomenon is extremely important since cleaning and disinfection techniques are usually largely dependent on the cleaning individuals and processes, and the use of biocides at an inappropriate concentration can lead not only to the overgrowth of resistant strains but to a further increase in the strains' resistance to disinfectants. However, the increased expression of the efflux pumps can induce resistance to various antibiotics through cross-resistance, representing an additional treatment failure and hospital outbreaks.

Appreciating this finding, it is very important to conclude that, in addition to antibiotic stewardship, the policy for disinfectant use should be implemented as part of infection-control activities. Responsible antibiotic policy and outbreak management must include a responsible disinfectant policy recognizing the spread of resistance.

Our study has inherent limitations. First, there was no implemented screening of healthcare workers for multi-resistant pathogens during the outbreak period. Second, no air sampling and analysis were carried out in the intensive care unit. Another limitation, if it can be called a limitation and not a strength, is that only a few vulnerable patients were affected by CRAB, so it was not possible to perform a statistical analysis of the risk factors. Furthermore, it would be worth examining the susceptibility and biofilm inhibitory effect of additional disinfectants, as in this study, only the effectiveness of the disinfectants used in the intensive care unit was investigated.

4. Materials and Methods

4.1. Study Design

In our retrospective cohort study, we described the epidemiological investigations and infection-control measures during a CRAB outbreak at the 175-bed Department of Neurosurgery and Neurointervention, Semmelweis University, Budapest, Hungary. The neurosurgical intensive care unit (ICU) comprises 18 beds with three5-bed rooms and three1-bed rooms. The patients housed here are with acute neurological disorders or after neurosurgical procedures. One of the 5-bed rooms is used for the 24 h observation of postoperative patients. Two hundred and ninety-one patients were treated together with postoperative patients in the ICU during the study period between 28 December 2023 and 22 February 2024.

Descriptive analysis summarized patient demographics, neurological diseases, surgical treatment, underlying diseases, infections, and outcomes.

4.2. Environmental Sampling

Environmental microbiological investigations were conducted in both the ICU and operating rooms in relation to the surgical procedures. Environmental samples were collected from frequently touched surfaces in the patient zone (e.g., respirators and monitors), shared areas (e.g., medication preparation area and storage), medical devices (e.g., anesthesia machine), and sinks (sink and sink drain) at the ICU and in the operating rooms as well.

Between December 2023 and February 2024, a total of 174 environmental samples were collected from the ICU and the operating rooms. Sampling was repeated several times.

4.3. Bacterial Strain

During the outbreak, six *A. baumannii* strains were isolated from the patients. All *A. baumannii* isolates were identified by MALDI-TOF/MS (Bruker Daltonik GmbH, Bremen, Germany). Antibacterial susceptibility testing was performed on all strains, and the results were interpreted according to EUCAST guidelines (www.eucast.org), accessed 1 January 2024.

The MALDI-TOF/MS system was not used for epidemiological purposes in the routine clinical microbiology laboratory. Rather, single-nucleotide polymorphism (SNP) analysis and epidemiological data from The National Center for Public Health and Pharmacy, Budapest, Hungary, revealed a putative outbreak clone comprising six CRAB strains belonging to the globally disseminated international clone ST2. These strains had no SNP differences and had identical antimicrobial resistance and virulence genes.

4.4. Minimal Inhibitory Concentration and Minimal Bactericidal Concentration Determination

In order to assess the effectiveness of different disinfectants, a microdilution assay was carried out. Two hundred µL of bacterial broth was added into each well of a 96-well plate. These broth wells were then treated with varying concentrations (ranging from 128 mg/mL to 0.06 µg/mL) of benzalkonium chloride and chlorhexidine digluconate. The wells were subsequently incubated for 24 h at 37 °C. After 24 h, the minimal inhibitory concentrations were determined. MIC endpoint is the lowest concentration of disinfectants where no visible growth is seen in the wells. After the MIC determination aliquots, wells showing no visible bacterial growth were seeded on BHI agar plates and incubated for 24 h at 37 °C. The lowest concentration of an antimicrobial agent, when 99.9% of the bacterial population is killed, is termed the MBC endpoint.

4.5. Killing Curve Assay

We made two-fold serial dilutions of the active compounds of different disinfectants in Müller–Hinton Broth, and 200 µL was transferred to a 96-well microtiter plate in duplicate. As positive control of the culture medium, Müller–Hinton Broth was used. AMcFarland 0.5 suspension was prepared in MH Broth from overnight colonies of *A. baumannii* cultured on Columbia Blood Agar Medium, and the suspension was later diluted 1:20 in MH Broth for optimal bacterial density. As an inoculum, 20 µL of this diluted suspension was added to each well, except for negative controls. The plate was then incubated at 37 °C for 24 h, and the optical density at 595 nm was recorded every 15 min. However, some antibacterial agents cause the elevation of the turbidity spontaneously, as on optical background for distraction, we used the MH Broth containing the corresponding concentration of every antibacterial compound. Negative controls and background controls were adjusted with 20 µL of MH Broth without bacteria.

4.6. Investigation of Biofilm Production

To detect the biofilm formation, we followed the protocol reported by Hassan et al. [43]. Bacteria were incubated in EMEM medium (Lonza Bioscience, Budapest, Hungary) at 37 °C for 24 h. The cultures were then diluted 1:100 with fresh medium. Individual wells of sterile flat-bottom 96-well polystyrene tissue culture plates (Biologix Ltd., Hallbergmoos, Germany) were filled with 200 µL of the diluted cultures. Control *A. baumannii* BAA-1805 was similarly incubated, diluted, and added to the tissue culture plate. Negative control wells only contained inoculated sterile broth. The plates were incubated at 37 °C for 48 h. After incubation, contents of each well were removed by gentle tapping. The wells were washed with 0.2 mL of phosphate-buffered saline (pH 7.2) four times. This treatment removed free-floating bacteria. Biofilm formed by bacteria adherent to the wells was fixed by 2% sodium acetate and stained by crystal violet (0.1%). Excess stain was removed by using deionized water, and plates were kept for drying. The optical density (OD) of the stained adherent biofilm was read on a Multiskan FC Microplate Photometer (Thermo Scientific, Budapest, Hungary) at 595 nm. The experiment was performed in triplicate and repeated three times. The interpretation of biofilm production followed the criteria of Stepanovic et al. [56]. To calculate classification breakpoints, the following breakpoints were set during our analyses: OD control = 0.153, non-biofilm producer: OD $\leq$ 0.153, weak biofilm producer: 0.306 $\geq$ OD > 0.153, medium biofilm producer: 0.617 $\geq$ OD > 0.306, and strong biofilm producer: OD > 0.617.

4.7. Minimum Biofilm Inhibitory Concentration (MBIC) Determination

To detect the minimum biofilm inhibitory concentration, the method described above was used with the following modifications. Bacteria were incubated in EMEM medium (Lonza Bioscience, Budapest, Hungary) at 37 °C for 24 h. Cultures were then diluted 1:100 with fresh medium. Individual wells of sterile flat bottom 96 well polystyrene tissue culture plates (Biologix Ltd.) were filled with 200 µL of the diluted cultures. The broth wells were then treated with varying concentrations (ranging from 128 mg/mL to 0.06 µg/mL) of serially diluted benzalkonium chloride and chlorhexidine digluconate. The plates were subsequently incubated for 24 h at 37 °C. Control *A. baumannii* BAA-1805 was also incubated, diluted, and added to the tissue culture plate. Negative control wells contained inoculated sterile broth. The plates were incubated at 37 °C for 48 h. After incubation, contents of each well were removed by gentle tapping. The wells were washed with 0.2 mL of phosphate-buffered saline (pH 7.2) four times. This procedure removed free-floating bacteria. Biofilm formed by bacteria adherent to the wells was fixed with 2% sodium acetate and stained with crystal violet (0.1%). Excess stain was removed with deionized water, and plates were kept for drying. The optical density (OD) of the stained adherent biofilm was read on a Multiskan FC Microplate Photometer (Thermo Scientific, Budapest, Hungary) at 595 nm. The experiment was performed in triplicate and repeated three times. The interpretation of biofilm production was performed according to Stepanovic et al. [56]. The lowest concentration of benzalkonium chloride, chlorhexidine digluconate necessary to inhibit bacterial biofilm formation was identified as the minimal biofilm inhibitory concentration.

4.8. Whole-Genome Sequencing (WGS)

WGS analysis was performed on one *A. baumannii* strain (MACI-KO) by using the Illumina MiSeq system in Eurofins BIOMI Kft. (Gödöllő, Hungary). Briefly, genomic DNA was extracted with the NucleoSpin Microbial DNA Mini kit (Macherey-Nagel, Düren, Germany). The amount of isolated DNA was measured by Qubit fluorometer, and the quality of DNA was tested by microcapillary electrophoresis (Tape Station 4150, Agilent, Waldbronn, Germany). Libraries were prepared with the Illumina DNA Prep kit according to the manufacturer's instructions. Sequencing was performed on an Illumina Miseq system using a MiSeq Reagent Kit v2, generating 250 bp paired-end reads. Genome assembly was performed with the SPAdes Genome assembler algorithm v. 3.15.3. [57]. Antibiotic-resistance genes were detected in the assembled genome using Bionumerics v. 8.1 software. The visualization was created with the Proksee server (https://proksee.ca/, accessed on 26 July 2024, London, UK) [58], and it is based on genome annotation data generated by RAST (https://rast.nmpdr.org/rast.cgi, accessed on 9 October 2020, Chicago, IL, USA) [59,60]. The assembled genome of *A. baumannii* strain was analyzed using the traditional 7-gene multilocus sequence typing (MLST).

4.9. Analysis of Disinfectant Resistance Genes by Quantitative PCR

The *A. baumannii* strain was co-cultivated with 0.0039% and 0.0078% concentrations of MBF at 0.0039% and 0.0078% concentrations of IP disinfecting agents in a 96-well microtiter plate at 37 °C. After overnight incubation, the total RNA of the bacteria was isolated with an innuPREP RNA Mini Kit 2.0 (Analytik Jena GmbH, Hardegsen, Germany) according to manufacturer's instructions. RNA concentrations were determined using a NanoDrop

1000 spectrophotometer (Thermo Fisher Scientific, Budapest, Hungary). Ten to one hundred ng of RNA was used for RT-PCR assay performed using the PrimeScript RT reagent kit (Takara Bio, Saint-Germain-en-Laye, France) and amplified the resulting cDNA on a qTOWER 3G (Analytik Jena GmbH, Hardegsen, Germany) instrument in the presence of selected primers.

The primers for *qacE1* genes were 5′-GTATTGGCGACCGCTTTTCTCG-3′ forward and 5′-AGCCGACTGTAATAAAACCAATCCC-3′reverse. The primers for *qacEΔ1* were 5′-TGCTTATGCAGTCTGGTCGGG-3′ forward and 5′-ACCTACAAAGCCCCACGCATC-3′ reverse. The primers for *16S* were 5′-CATGCCGCGTGTGTGAAGAAG-3′ forward and 5′-AGCCGGTGCTTATTCTGCGAG-3′ reverse. Relative mRNA expression was calculated by means of the change in cycle threshold (ΔΔCt) and normalized to the geometric mean of housekeeping gene 16S. Basal mRNA levels of QacE1 and QacEΔ1 were compared with those of housekeeping gene 16S by calculating the difference between their Ct.

5. Conclusions

Here, we demonstrate that a complex infection control approach is important to effectively control the spread of nosocomial pathogens during an outbreak. We highlight the role of routinely used disinfectant agents during the cleaning procedures. Our work also highlights that considering disinfectant tolerance and resistance is important in the control of hospital epidemics. We recommend setting up phenotypic disinfectant susceptibility test standardization for outbreak strains causing hospital outbreaks. Whole-genome sequencing analysis of the outbreak strains is also useful. Or, for laboratories with limited resources, at least PCR-based rapid detection of *qacE1* and *qacEΔ1*, widespread mobile disinfectant resistance genes of Gram-negative HAI pathogens, would be desirable.

Author Contributions: J.K. (Jozsef Kelemen): conceptualization, writing—original draft, M.S.: data curation, investigation, E.K.D.: data curation, project administration, J.B.: supervision, J.K. (Jozsef Katona): investigation, Z.S.: validation, L.S.: methodology, supervision, writing—review editing, Z.P.: methodology, supervision, B.S.: resources, Z.A.D.: resources, project administration, J.J.: software, visualization, B.K.: software, formal analysis, F.M.: methodology, Z.D.: methodology, J.D.: data curation, investigation, D.S.: conceptualization, funding acquisition, supervision, writing—original draft, L.E.: conceptualization, supervision. All authors have read and agreed to the published version of the manuscript.

Funding: This study was supported by HUN-REN-SU, the Human Microbiota Research Group No. "0272", and by the European Union's Horizon 2020 research and innovation program (952491-AmReSu). B.K. was supported by the János Bolyai Scholarship (BO/00286/22/5) of the Hungarian Academy of Sciences.

Institutional Review Board Statement: Ethical approval was not required, as the isolates were collected as part of routine clinical care and patient data collection followed patient discharge from the hospital and/or death. No additional isolates were collected beyond those obtained from routine clinical care, and no diagnostic or treatment decisions were affected by the outcomes of this study.

Informed Consent Statement: Patient consent was waived because this is a retrospective observational study.

Data Availability Statement: The genome sequence of the *A. baumannii* strain (MACI-KO) is deposited in the NCBI database at the following bioproject accession number: PRJNA1130157.

Acknowledgments: We thank Akos Toth (the National Center for Public Health and Pharmacy, Budapest, Hungary) for providing data on single-nucleotide polymorphism (SNP) analysis and epidemiological information about CRAB outbreak strains. Akos Toth provided us with an official clinical report about the SNP data of all outbreak strains for the Department of Neurosurgery and Neurointervention, Semmelweis University, Budapest, Hungary, and gave us permission to use the results in this study.

Conflicts of Interest: The authors declare no conflicts of interest.

References

1. Obenhuber, T.; Scheier, T.C.; Stutz, T.; Hug, M.; Fontein, D.; Kaiser, A.; Schoene, S.; Steiger, P.; Brugger, S.D.; Zingg, W.; et al. An outbreak of multi-drug-resistant Acinetobacter baumannii on a burns ICU and its control with multi-faceted containment measures. *J. Hosp. Infect.* **2024**, *146*, 102–108. [CrossRef] [PubMed]
2. Baleivanualala, S.C.; Matanitobua, S.; Soqo, V.; Smita, S.; Limaono, J.; Sharma, S.C.; Devi, S.V.; Boseiwaqa, L.V.; Vera, N.; Kumar, S.; et al. Molecular and clinical epidemiology of carbapenem resistant Acinetobacter baumannii, Pseudomonas aeruginosa and Enterobacterales in Fiji: A multicentre prospective observational study. *Lancet Reg. Health West Pac.* **2024**, *47*, 101095. [CrossRef] [PubMed]
3. Su, P.W.; Yang, E.C.; Moi, S.H.; Yang, C.H.; Chuang, L.Y. Prevalence of Carbapenem Resistance Genes among Acinetobacter baumannii Isolated from a Teaching Hospital in Taiwan. *Antibiotics* **2023**, *12*, 1357. [CrossRef]
4. Tickler, I.A.; Kawa, D.; Obradovich, A.E.; Fang, F.C.; Tenover, F.C.; The Healthcare Associated Infections Consortium. Characterization of Carbapenemase- and ESBL-Producing Gram-Negative Bacilli Isolated from Patients with Urinary Tract and Bloodstream Infections. *Antibiotics* **2023**, *12*, 1386. [CrossRef]
5. Seifert, H.; Muller, C.; Stefanik, D.; Higgins, P.G.; Wohlfarth, E.; Kresken, M. In Vitro Activity of Cefiderocol against a Global Collection of Carbapenem-Resistant Acinetobacter baumannii Isolates. *Antibiotics* **2023**, *12*, 1172. [CrossRef]
6. Ruekit, S.; Srijan, A.; Serichantalergs, O.; Margulieux, K.R.; Mc Gann, P.; Mills, E.G.; Stribling, W.C.; Pimsawat, T.; Kormanee, R.; Nakornchai, S.; et al. Molecular characterization of multidrug-resistant ESKAPEE pathogens from clinical samples in Chonburi, Thailand (2017–2018). *BMC Infect. Dis.* **2022**, *22*, 695. [CrossRef]
7. Piperaki, E.T.; Tzouvelekis, L.S.; Miriagou, V.; Daikos, G.L. Carbapenem-resistant Acinetobacter baumannii: In pursuit of an effective treatment. *Clin. Microbiol. Infect.* **2019**, *25*, 951–957. [CrossRef]
8. Bonomo, R.A.; Szabo, D. Mechanisms of multidrug resistance in Acinetobacter species and Pseudomonas aeruginosa. *Clin. Infect. Dis.* **2006**, *43* (Suppl. S2), S49–S56. [CrossRef]
9. Zhao, S.; Zhang, B.; Liu, C.; Sun, X.; Chu, Y. Acinetobacter baumannii infection in intensive care unit: Analysis of distribution and drug resistance. *Mol. Biol. Rep.* **2024**, *51*, 120. [CrossRef]
10. Paul, M.; Carrara, E.; Retamar, P.; Tangden, T.; Bitterman, R.; Bonomo, R.A.; de Waele, J.; Daikos, G.L.; Akova, M.; Harbarth, S.; et al. European Society of Clinical Microbiology and Infectious Diseases (ESCMID) guidelines for the treatment of infections caused by multidrug-resistant Gram-negative bacilli (endorsed by European society of intensive care medicine). *Clin. Microbiol. Infect.* **2022**, *28*, 521–547. [CrossRef]
11. Tacconelli, E.; Carrara, E.; Savoldi, A.; Harbarth, S.; Mendelson, M.; Monnet, D.L.; Pulcini, C.; Kahlmeter, G.; Kluytmans, J.; Carmeli, Y.; et al. Discovery, research, and development of new antibiotics: The WHO priority list of antibiotic-resistant bacteria and tuberculosis. *Lancet Infect. Dis.* **2018**, *18*, 318–327. [CrossRef] [PubMed]
12. Miller, W.R.; Arias, C.A. ESKAPE pathogens: Antimicrobial resistance, epidemiology, clinical impact and therapeutics. *Nat. Rev. Microbiol.* **2024**, *22*, 598–616. [CrossRef] [PubMed]
13. Tian, C.; Di, L.; Dong, S.; Tian, X.; Huang, D.; Zhao, Y.; Chen, J.; Xia, D.; Wang, S. Whole genome sequencing and genomic characteristics analysis of carbapenem-resistant Acinetobacter baumannii clinical isolates in two hospitals in China. *Infect. Genet. Evol.* **2024**, *123*, 105642. [CrossRef] [PubMed]
14. Jauneikaite, E.; Baker, K.S.; Nunn, J.G.; Midega, J.T.; Hsu, L.Y.; Singh, S.R.; Halpin, A.L.; Hopkins, K.L.; Price, J.R.; Srikantiah, P.; et al. Genomics for antimicrobial resistance surveillance to support infection prevention and control in health-care facilities. *Lancet Microbe* **2023**, *4*, e1040–e1046. [CrossRef] [PubMed]
15. Wong, D.; Nielsen, T.B.; Bonomo, R.A.; Pantapalangkoor, P.; Luna, B.; Spellberg, B. Clinical and Pathophysiological Overview of Acinetobacter Infections: A Century of Challenges. *Clin. Microbiol. Rev.* **2017**, *30*, 409–447. [CrossRef] [PubMed]
16. Hajihashemi, B.; Abbasi, A.; Shokri, D. Emergence of colistin resistant Acinetobacter baumannii clonal complex 2 (CC2) among hospitalized patients in Iran. *Acta Microbiol. Immunol. Hung.* **2023**, *70*, 213–219. [CrossRef]
17. Seifert, H.; Blondeau, J.; Lucassen, K.; Utt, E.A. Global update on the in vitro activity of tigecycline and comparators against isolates of Acinetobacter baumannii and rates of resistant phenotypes (2016–2018). *J. Glob. Antimicrob. Resist.* **2022**, *31*, 82–89. [CrossRef]
18. Attia, N.M.; Elbaradei, A. Fluoroquinolone resistance conferred by gyrA, parC mutations, and AbaQ efflux pump among Acinetobacter baumannii clinical isolates causing ventilator-associated pneumonia. *Acta Microbiol. Immunol. Hung.* **2019**, *67*, 234–238. [CrossRef]
19. Namaki Kheljan, M.; Hassanzadeh, M.; Srdari Jabedar, M.; Mohammadi Gollou, A.; Ashouri, P.; Teimourpour, R.; Arzanlou, M. Characterization of disinfectant susceptibility profiles among clinical isolates of Acinetobacter baumannii in Ardabil, Iran. *Acta Microbiol. Immunol. Hung.* **2023**, *70*, 311–317. [CrossRef]
20. McDonnell, G.; Russell, A.D. Antiseptics and disinfectants: Activity, action, and resistance. *Clin. Microbiol. Rev.* **1999**, *12*, 147–179. [CrossRef]
21. Boyce, J.M. Quaternary ammonium disinfectants and antiseptics: Tolerance, resistance and potential impact on antibiotic resistance. *Antimicrob. Resist. Infect. Control* **2023**, *12*, 32. [CrossRef]
22. Maillard, J.Y.; Centeleghe, I. How biofilm changes our understanding of cleaning and disinfection. *Antimicrob. Resist. Infect. Control* **2023**, *12*, 95. [CrossRef]

23. Amodeo, D.; Manzi, P.; De Palma, I.; Puccio, A.; Nante, N.; Barcaccia, M.; Marini, D.; Pietrella, D. Efficacy of Violet-Blue (405 nm) LED Lamps for Disinfection of High-Environmental-Contact Surfaces in Healthcare Facilities: Leading to the Inactvation of Microorganisms and Reduction of MRSA Contamination. *Pathogens* **2023**, *12*, 1338. [CrossRef]
24. Messina, G.; Rosadini, D.; Burgassi, S.; Messina, D.; Nante, N.; Tani, M.; Cevenini, G. Tanning the bugs—A pilot study of an innovative approach to stethoscope disinfection. *J. Hosp. Infect.* **2017**, *95*, 228–230. [CrossRef]
25. Tanguy, M.; Kouatchet, A.; Tanguy, B.; Pichard, É.; Fanello, S.; Joly-Guillou, M.L. Management of an Acinetobacter baumannii outbreak in an intensive care unit. *Med. Mal. Infect.* **2017**, *47*, 409–414. [CrossRef]
26. Gramatniece, A.; Silamikelis, I.; Zahare, I.; Urtans, V.; Zahare, I.; Dimina, E.; Saule, M.; Balode, A.; Radovica-Spalvina, I.; Klovins, J.; et al. Control of Acinetobacter baumannii outbreak in the neonatal intensive care unit in Latvia: Whole-genome sequencing powered investigation and closure of the ward. *Antimicrob. Resist. Infect. Control* **2019**, *8*, 84. [CrossRef]
27. Ostorhazi, E.; Rozgonyi, F.; Sztodola, A.; Harmos, F.; Kovalszky, I.; Szabo, D.; Knappe, D.; Hoffmann, R.; Cassone, M.; Wade, J.D.; et al. Preclinical advantages of intramuscularly administered peptide A3-APO over existing therapies in Acinetobacter baumannii wound infections. *J. Antimicrob. Chemother.* **2010**, *65*, 2416–2422. [CrossRef]
28. Dancer, S.J. Controlling hospital-acquired infection: Focus on the role of the environment and new technologies for decontamination. *Clin. Microbiol. Rev.* **2014**, *27*, 665–690. [CrossRef]
29. Cassini, A.; Hogberg, L.D.; Plachouras, D.; Quattrocchi, A.; Hoxha, A.; Simonsen, G.S.; Colomb-Cotinat, M.; Kretzschmar, M.E.; Devleesschauwer, B.; Cecchini, M.; et al. Attributable deaths and disability-adjusted life-years caused by infections with antibiotic-resistant bacteria in the EU and the European Economic Area in 2015: A population-level modelling analysis. *Lancet Infect. Dis.* **2019**, *19*, 56–66. [CrossRef]
30. Landman, D.; Quale, J.M.; Mayorga, D.; Adedeji, A.; Vangala, K.; Ravishankar, J.; Flores, C.; Brooks, S. Citywide clonal outbreak of multiresistant Acinetobacter baumannii and Pseudomonas aeruginosa in Brooklyn, NY: The preantibiotic era has returned. *Arch. Intern. Med.* **2002**, *162*, 1515–1520. [CrossRef]
31. Firoozeh, F.; Bakhshi, F.; Dadashi, M.; Badmasti, F.; Zibaei, M.; Omidinia, N. Detection of multidrug-resistant Acinetobacter baumannii from burn patients and healthcare workers in Iran. *Acta Microbiol. Immunol. Hung.* **2023**, *70*, 22–28. [CrossRef]
32. Weinberg, S.E.; Villedieu, A.; Bagdasarian, N.; Karah, N.; Teare, L.; Elamin, W.F. Control and management of multidrug resistant Acinetobacter baumannii: A review of the evidence and proposal of novel approaches. *Infect. Prev. Pract.* **2020**, *2*, 100077. [CrossRef]
33. Dijkshoorn, L.; Nemec, A.; Seifert, H. An increasing threat in hospitals: Multidrug-resistant Acinetobacter baumannii. *Nat. Rev. Microbiol.* **2007**, *5*, 939–951. [CrossRef]
34. Tacconelli, E.; Cataldo, M.A.; Dancer, S.J.; De Angelis, G.; Falcone, M.; Frank, U.; Kahlmeter, G.; Pan, A.; Petrosillo, N.; Rodriguez-Bano, J.; et al. ESCMID guidelines for the management of the infection control measures to reduce transmission of multidrug-resistant Gram-negative bacteria in hospitalized patients. *Clin. Microbiol. Infect.* **2014**, *20* (Suppl. S1), 1–55. [CrossRef]
35. World Health Organization. *Guidelines for the Prevention and Control of Carbapenem-Resistant Enterobacteriaceae, Acinetobacter baumannii and Pseudomonas aeruginosa in Health Care Facilities*; World Health Organization: Geneva, Switzerland, 2017; Available online: https://apps.who.int/iris/handle/10665/259462 (accessed on 27 May 2022).
36. Tacconelli, E.; Buhl, M.; Humphreys, H.; Malek, V.; Presterl, E.; Rodriguez-Bano, J.; Vos, M.C.; Zingg, W.; Mutters, N.T. Analysis of the challenges in implementing guidelines to prevent the spread of multidrug-resistant gram-negatives in Europe. *BMJ Open* **2019**, *9*, e027683. [CrossRef]
37. Ben-Chetrit, E.; Wiener-Well, Y.; Lesho, E.; Kopuit, P.; Broyer, C.; Bier, L.; Assous, M.V.; Benenson, S.; Cohen, M.J.; McGann, P.T.; et al. An intervention to control an ICU outbreak of carbapenem-resistant Acinetobacter baumannii: Long-term impact for the ICU and hospital. *Crit. Care* **2018**, *22*, 319. [CrossRef]
38. Gray, A.P.; Allard, R.; Pare, R.; Tannenbaum, T.; Lefebvre, B.; Levesque, S.; Mulvey, M.; Maalouf, L.; Perna, S.; Longtin, Y. Management of a hospital outbreak of extensively drug-resistant Acinetobacter baumannii using a multimodal intervention including daily chlorhexidine baths. *J. Hosp. Infect.* **2016**, *93*, 29–34. [CrossRef]
39. Meschiari, M.; Lopez-Lozano, J.M.; Di Pilato, V.; Gimenez-Esparza, C.; Vecchi, E.; Bacca, E.; Orlando, G.; Franceschini, E.; Sarti, M.; Pecorari, M.; et al. A five-component infection control bundle to permanently eliminate a carbapenem-resistant Acinetobacter baumannii spreading in an intensive care unit. *Antimicrob. Resist. Infect. Control* **2021**, *10*, 123. [CrossRef]
40. Metan, G.; Zarakolu, P.; Otlu, B.; Tekin, I.; Aytac, H.; Bolek, E.C.; Metin, B.C.; Arsava, E.M.; Unal, S. Emergence of colistin and carbapenem-resistant Acinetobacter calcoaceticus-Acinetobacter baumannii (CCR-Acb) complex in a neurological intensive care unit followed by successful control of the outbreak. *J. Infect. Public Health* **2020**, *13*, 564–570. [CrossRef]
41. Valencia-Martin, R.; Gonzalez-Galan, V.; Alvarez-Marin, R.; Cazalla-Foncueva, A.M.; Aldabo, T.; Gil-Navarro, M.V.; Alonso-Araujo, I.; Martin, C.; Gordon, R.; Garcia-Nunez, E.J.; et al. A multimodal intervention program to control a long-term Acinetobacter baumannii endemic in a tertiary care hospital. *Antimicrob. Resist. Infect. Control* **2019**, *8*, 199. [CrossRef]
42. Chung, Y.K.; Kim, J.S.; Lee, S.S.; Lee, J.A.; Kim, H.S.; Shin, K.S.; Park, E.Y.; Kang, B.S.; Lee, H.J.; Kang, H.J. Effect of daily chlorhexidine bathing on acquisition of carbapenem-resistant Acinetobacter baumannii (CRAB) in the medical intensive care unit with CRAB endemicity. *Am. J. Infect. Control* **2015**, *43*, 1171–1177. [CrossRef]
43. Karampatakis, T.; Tsergouli, K.; Iosifidis, E.; Antachopoulos, C.; Karapanagiotou, A.; Karyoti, A.; Gritsi-Gerogianni, N.; Tsakris, A.; Roilides, E. Impact of active surveillance and infection control measures on carbapenem-resistant Gram-negative bacterial colonization and infections in intensive care. *J. Hosp. Infect.* **2018**, *99*, 396–404. [CrossRef]

44. Fan, C.Y.; Lee, W.T.; Hsu, T.C.; Lee, C.H.; Wang, S.P.; Chen, W.S.; Huang, C.H.; Lee, C.C. Effect of chlorhexidine bathing on colonization or infection with Acinetobacter baumannii: A systematic review and meta-analysis. *J. Hosp. Infect.* **2019**, *103*, 284–292. [CrossRef]
45. Enfield, K.B.; Huq, N.N.; Gosseling, M.F.; Low, D.J.; Hazen, K.C.; Toney, D.M.; Slitt, G.; Zapata, H.J.; Cox, H.L.; Lewis, J.D.; et al. Control of Simultaneous Outbreaks of Carbapenemase-Producing Enterobacteriaceae and Extensively Drug-ResistantAcinetobacter baumannii Infection in an Intensive Care Unit Using Interventions Promoted in the Centers for Disease Control and Prevention 2012 Carbapenemase-Resistant Enterobacteriaceae Toolkit. *Infect. Control Hosp. Epidemiol.* **2014**, *35*, 810–817.
46. Eckardt, P.; Canavan, K.; Guran, R.; George, E.; Miller, N.; Avendano, D.H.; Kim, M.; Himed, K.; Ramirez, K.H.G. Containment of a carbapenem-resistant Acinetobacter baumannii complex outbreak in a COVID-19 intensive care unit. *Am. J. Infect. Control* **2022**, *50*, 477–481. [CrossRef]
47. Lerner, A.O.; Abu-Hanna, J.; Carmeli, Y.; Schechner, V. Environmental contamination by carbapenem-resistant Acinetobacter baumannii: The effects of room type and cleaning methods. *Infect. Control Hosp. Epidemiol.* **2020**, *41*, 166–171. [CrossRef]
48. Harris, A.D.; Johnson, J.K.; Pineles, L.; O'Hara, L.M.; Bonomo, R.A.; Thom, K.A. Patient-to-Patient Transmission of Acinetobacter baumannii Gastrointestinal Colonization in the Intensive Care Unit. *Antimicrob. Agents Chemother.* **2019**, *63*, e00392-19. [CrossRef]
49. Cho, O.H.; Bak, M.H.; Baek, E.H.; Park, K.-H.; Kim, S.; Bae, I.G. Successful control of carbapenem-resistant Acinetobacter baumannii in a Korean university hospital: A 6-year perspective. *Am. J. Infect. Control* **2014**, *42*, 976–979. [CrossRef]
50. Perez, S.; Innes, G.K.; Walters, M.; Mehr, J.; Arias, J.; Greeley, R.; Chew, D. Increase in Hospital-Acquired Carbapenem-Resistant Acinetobacter baumannii Infection and Colonization in an Acute Care Hospital during a Surge in COVID-19 Admissions—New Jersey, February–July 2020. MMWR. *Morb. Mortal. Wkly. Rep.* **2020**, *69*, 1827–1831. [CrossRef]
51. Munoz-Price, L.S.; Carling, P.; Cleary, T.; Fajardo-Aquino, Y.; DePascale, D.; Jimenez, A.; Hughes, M.; Namias, N.; Pizano, L.; Kett, D.H.; et al. Control of a two-decade endemic situation with carbapenem-resistant Acinetobacter baumannii: Electronic dissemination of a bundle of interventions. *Am. J. Infect. Control* **2014**, *42*, 466–471. [CrossRef]
52. Medioli, F.; Bacca, E.; Faltoni, M.; Burastero, G.J.; Volpi, S.; Menozzi, M.; Orlando, G.; Bedini, A.; Franceschini, E.; Mussini, C.; et al. Is It Possible to Eradicate Carbapenem-Resistant Acinetobacter baumannii (CRAB) from Endemic Hospitals? *Antibiotics* **2022**, *11*, 1015. [CrossRef]
53. Gomaa, F.A.M.; Helal, Z.H.; Khan, M.I. High Prevalence of blaNDM-1, blaVIM, qacE, and qacEΔ1 genes and their association with decreased susceptibility to antibiotics and common hospital biocides in clinical isolates of Acinetobacter baumannii. *Microorganisms* **2017**, *5*, 18. [CrossRef]
54. Elkhatib, W.F.; Khalil, M.A.F.; Ashour, H.M. Integrons and Antiseptic Resistance Genes Mediate Resistance of Acinetobacter baumannii and Pseudomonas aeruginosa Isolates from Intensive Care Unit Patients with Wound Infections. *Curr. Mol. Med.* **2019**, *19*, 286–293. [CrossRef]
55. Slipski, C.J.; Jamieson-Datzkiw, T.R.; Zhanel, G.G.; Bay, D.C. Characterization of Proteobacterial Plasmid Integron-Encoded qac Efflux Pump Sequence Diversity and Quaternary Ammonium Compound Antiseptic Selection in Escherichia coli Grown Planktonically and as Biofilms. *Antimicrob. Agents Chemother.* **2021**, *65*, e0106921. [CrossRef]
56. Stepanovic, S.; Vukovic, D.; Hola, V.; Di Bonaventura, G.; Djukic, S.; Cirkovic, I.; Ruzicka, F. Quantification of biofilm in microtiter plates: Overview of testing conditions and practical recommendations for assessment of biofilm production by staphylococci. *APMIS* **2007**, *115*, 891–899. [CrossRef]
57. Overbeek, R.; Olson, R.; Pusch, G.D.; Olsen, G.J.; Davis, J.J.; Disz, T.; Edwards, R.A.; Gerdes, S.; Parrello, B.; Shukla, M.; et al. The SEED and the Rapid Annotation of microbial genomes using Subsystems Technology (RAST). *Nucleic Acids Res.* **2014**, *42*, D206–D214. [CrossRef]
58. Grant, J.R.; Enns, E.; Marinier, E.; Mandal, A.; Herman, E.K.; Chen, C.Y.; Graham, M.; Van Domselaar, G.; Stothard, P. Proksee: In-depth characterization and visualization of bacterial genomes. *Nucleic Acids Res.* **2023**, *51*, W484–W492. [CrossRef]
59. Bankevich, A.; Nurk, S.; Antipov, D.; Gurevich, A.A.; Dvorkin, M.; Kulikov, A.S.; Lesin, V.M.; Nikolenko, S.I.; Pham, S.; Prjibelski, A.D.; et al. SPAdes: A new genome assembly algorithm and its applications to single-cell sequencing. *J. Comput. Biol.* **2012**, *19*, 455–477. [CrossRef]
60. Aziz, R.K.; Bartels, D.; Best, A.A.; DeJongh, M.; Disz, T.; Edwards, R.A.; Formsma, K.; Gerdes, S.; Glass, E.M.; Kubal, M.; et al. The RAST Server: Rapid annotations using subsystems technology. *BMC Genom.* **2008**, *9*, 75. [CrossRef]

antibiotics

MDPI

Article

Co-Administration of High-Dose Nebulized Colistin for *Acinetobacter baumannii* Bacteremic Ventilator-Associated Pneumonia: Impact on Outcomes

Ioannis Andrianopoulos [1], Nikolaos Kazakos [1], Nikolaos Lagos [1], Theodora Maniatopoulou [1], Athanasios Papathanasiou [1], Georgios Papathanakos [1], Despoina Koulenti [2,3,*], Eleni Toli [1], Konstantina Gartzonika [4] and Vasilios Koulouras [1]

1 Intensive Care Unit, University Hospital of Ioannina, 45500 Ioannina, Greece; jandri0@yahoo.gr (I.A.); nikazakos@yahoo.com (N.K.); lagman75@gmail.com (N.L.); doramaniatopoulou@yahoo.gr (T.M.); thanasis.papathanasiou@gmail.com (A.P.); gppthan@icloud.com (G.P.); eletoli87@gmail.com (E.T.); vpkoulouras@yahoo.gr (V.K.)
2 UQ Centre for Clinical Research, Faculty of Medicine, The University of Queensland, Brisbane, QLD 4029, Australia
3 Second Critical Care Department, Attikon University Hospital, Rimini Street, 12462 Athens, Greece
4 Department of Microbiology, University Hospital of Ioannina, 45500 Ioannina, Greece; kgartzon@uoi.gr
* Correspondence: d.koulenti@uq.edu.au or despoina.koulenti@nhs.net

Abstract: *Acinetobacter baumannii* (*A. baumannii*) is a difficult-to-treat (DTR) pathogen that causes ventilator-associated pneumonia (VAP) associated with high mortality. To improve the outcome of DTR A. *Baumannii* VAP, nebulized colistin (NC) was introduced with promising but conflicting results on mortality in earlier studies. Currently, NC is used at a much higher daily dose compared to the past. Nevertheless, there is little evidence on the effect of high-dose NC on the outcomes of *A.* baumannii VAPs, especially in the current era where the percentage of colistin-resistant *A. baumannii* strains is rising. We conducted a retrospective study comparing bacteremic *A. baumannii* VAP patients who were treated with and without NC co-administration and were admitted in the Intensive Care Unit of University Hospital of Ioannina from March 2020 to August 2023. Overall, 59 patients (21 and 38 with and without NC coadministration, respectively) were included. Both 28-day and 7-day mortalities were significantly lower in the patient group treated with NC (52.4% vs. 78.9%, *p* 0.034 and 9.5% vs. 47.4%, *p* 0.003, respectively). Patients treated with NC had a higher percentage of sepsis resolution by day 7 (38.1% vs. 13.5%, *p* 0.023) and were more likely to be off vasopressors by day 7 (28.6% vs. 8.1%, *p* 0.039). The addition of NC in the treatment regime of *A. baumannii* VAP decreased mortality.

Keywords: *Acinetobacter baumannii*; bacteremia; ventilator-associated pneumonia; nebulized colistin; mortality; critically ill

1. Introduction

Acinetobacter baumannii (*A. baumannii*) is a multi-drug-resistant (MDR) pathogen associated with severe infections, in particular, primary bacteremias, pneumonia and urinary tract infections [1,2]. It is an established difficult-to-treat (DTR) pathogen and its prevalence among hospital infections is rising [3]. Therefore, it is now being considered by the WHO as a critical priority to identify novel treatments for [4]. Mortality in such infections is particularly high among patients with bacteremias, pneumonia or sepsis [5–8]. It appears that critically ill patients that develop *A. baumannii* bacteremia due to ventilator-associated pneumonia (VAP) have worse prognosis, especially when they develop septic shock as a complication [9]. And it seems that the emergence of the COVID-19 pandemic has caused a rise in the prevalence of bacterial superinfections from MDR and DTR pathogens [10,11], as well as in the prevalence of VAP as a cause of bacteremia in critically ill patients [12].

Part of the reason for the *A. baumannii* VAP-associated high mortality is the fact that *A. baumannii* VAP infections are difficult to treat as therapeutic antibiotic choices are rather limited. In addition, penetration of intravenous antibiotics in the inflamed lung tissue is rather poor for most antibiotics, e.g., colistin b-lactams [13,14]. An attempt to overcome the poor penetration is to increase the dose or reduce the time interval between doses [15], though in cases of higher MIC, even that may not be adequate [16,17]. Therefore, to overcome this, especially in cases of pan-drug-resistant (PDR) pathogens, physicians have pursued combinations of antibiotics, although with conflicting results [18–23].

Another strategy to overcome the poor lung tissue penetration in cases of VAP is by using nebulized antimicrobials. Nebulized antimicrobials' theoretical advantage over the intravenous route is that nebulized antibiotics are delivered directly in the lung alveoli, and both experimental and clinical studies have shown that they offer far better lung tissue concentrations of important antibiotics such as aminoglycosides and colistin compared to the intravenous (IV) route [24,25]. Nebulized colistin, in particular, has been the subject of investigation in VAP from MDR pathogens in a few studies, with conflicting results on its effect on patient outcome. This inconsistency in results has been attributed to the differences in study population, dose of nebulized colistin, co-administration of IV colistin or other intravenous antibiotics and the use of different types of nebulizers [26,27]. In the most recently published meta-analysis, nebulized colistin showed a higher eradication rate but failed to show a statistically significant mortality rate improvement [26]. Nevertheless, there was significant heterogeneity among studies, and there is a need for further trials. This necessity for new trials is highlighted first of all by the fact that currently much higher doses (12–15 million units of colistin per day) than those studied in these earlier studies are used, and the effect of these higher doses on patient outcome is understudied [27].

In addition, as the percentage of DTR and colistin-resistant strains is rising, therapeutic choices become very limited, especially in cases of PDR strains or infections where it is hard to achieve adequate antibiotic concentrations, e.g., VAP and meningitis. In such cases, physicians end up using empirically used combinations of antibiotics, while evidence on the best combination remains scarce [28,29]. In PDR *A. baumannii* VAP cases, the combination of colistin, tigecycline and high dose sulbactam has showed promising results [30]. Interestingly, other studies also suggest that colistin appears to be the cornerstone of treatment, irrespective of its sensitivity status, especially when cefiderocol is unavailable as is the case in Greece [22]. If such is the case, then colistin is an essential drug in DTR *A. baumannii* VAP, and therefore, the hypothesis of improving treatment outcomes using adjunct nebulized colistin in addition to IV is raised.

The purpose of our study was to assess the effect of nebulized colistin in addition to the intravenous antibiotics on the mortality and treatment outcome of bacteremic DTR *A. baumannii* VAP.

2. Results

Overall, from 144 patients with bacteremia from *A. baumannii*, we identified 70 (48.6%) patients that had VAP as the source of their bacteremia. After excluding patients that received less than 24 h of antibiotics and patients that developed brain death, our analysis included 59 patients (44 male, 15 female) of mean age of 69 ± 9.6 years old. Out of these 59 patients, 21 received nebulized colistin as part of their antibiotic treatment regime (Group A) and 38 did not receive nebulized colistin as part of their antibiotic treatment regime (Group B). Eighteen (18) patients received 5 million units of nebulized colistin every 8 h, 1 patient received 6 million units of nebulized colistin every 8 h, and 2 patients received 3 million units every 8 h.

Baseline characteristics of the two groups of patients with VAP and associated bacteremia due to *A. baumannii* are shown in Table 1. Overall, there was no statistically significant difference in the baseline characteristics between the subgroup of patients that received combination of IV antibiotics and nebulized colistin (Group A) and the subgroup that received only IV antibiotics (Group B). Patients from Group A and B were of similar age (67.5 ± 9.8 and 67.3 ± 9.6, respectively, ns) and had similar SOFA (3 vs. 3, respectively, ns) and similar APACHE II (19 vs. 20, respectively, ns) scores. There was also no statistically significant difference in comorbidities. Regarding antimicrobial resistance, although Group A patients had a tendency for more resistant strains (a higher percentage of PDR, 28.6% vs. 13.2%, and a higher percentage of colistin resistance, 91.5% vs. 71.1%), altogether the proportion of PDR and colistin-resistant *A. baumannii* strains did not statistically differ between the two study groups. The great majority of patients were treated with a combination of antibiotics, while 33.3% of Group A patients and 31.6% of Group B patients received a triple combination of antibiotics (no statistical difference). Finally, there was no statistical difference between the two groups regarding the use of intravenous colistin (90.5% Group A vs. 78.9% in Group B, $p = 0.258$).

Table 1. Baseline characteristics of the patients with bacteremic VAP from *Acinetobacter baumannii* that were treated (Group A) and not treated (Group B) with nebulized colistin.

Parameter	Group A (Treated with Nebulized Colistin) (n = 21)	Group B (Treated without Nebulized) Colistin (n = 38)	p-Value
Age (mean ± SD)	67.5 ± 9.8	67.3 ± 9.6	0.657
Gender (male n, %)	15 (71.4%)	29 (76.3%)	0.680
APACHE II median (min–max)	19 (12–35)	20 (12–47)	0.934
CCI ∞ median (min–max)	3 (1–9)	3 (0–8)	0.898
SOFA ∞ score median (min–max)	4 (3–11)	4 (2–17)	0.798
COVID 19 (n, %)	19 (90.5%)	34 (89.5%)	0.903
Comorbidities			
Diabetes mellitus (n, %)	10 (47.6%)	14 (36.8%)	0.420
Heart failure (n, %)	1 (4.8%)	2 (5.3%)	0.933
Coronary Heart Disease (n, %)	3 (14.3%)	3 (7.9%)	0.437
Chronic Kidney Disease (n, %)	1 (4.8%)	3 (7.9%)	0.647
Cirrhosis (n, %)	1 (4.8%)	0 (0%)	0.175
COPD ∞ (n, %)	3 (14.3%)	5 (13.2%)	0.904
Obesity (n, %)	4 (19%)	5 (13.2%)	0.547
Cancer (n, %)	0(0%)	(4 10.5%)	0.124
Infection parameters			
PDR ∞	6 (28.6%)	5 (13.2%)	0.146
Colistin sensitive	2 (9.5%)	11 (28.9%)	0.085
Colistin treated	19 (90.5%)	30 (78.9%)	0.258
Patient was treated with:			
Monotherapy	0 (0)	2 (5.3%)	0.443
2 drugs	10 (47.6%)	21 (55.3%)	
3 drugs	7 (33.3%)	12 (31.6%)	
4 drugs	4 (19%)	3 (7.9%)	
Antibiotics			
Tigecycline (n, %)	21 (100%)	37 (97.4%)-	1
High dose ampicillin/sulbactam (n, %)	13 (62%)	8 (28.1%)	0.064

Table 1. *Cont.*

Parameter	Group A (Treated with Nebulized Colistin) (n = 21)	Group B (Treated without Nebulized) Colistin (n = 38)	p-Value
Meropenem (n, %)	7 (33.3%)	15 (39.5%)	0.781
Fosfomycin (n, %)	3 (14.3%)	4 (10.5%)	0.69
Appropriate antibiotic treatment in shock ʃ	17 (85%)	35 (97.2%)	0.089

∞ CCI: Charlson Comorbidity Index; SOFA, Sequential Organ Failure Assessment score; COPD, Chronic Obstructive Pulmonary Disease; PDR, pan-drug-resistant. ʃ The patient at the time of developing septic shock was already on or was started on antibiotics targeting *A. baumannii*.

The outcomes of both patient groups are shown in Table 2. Overall, the 28-day mortality was 69.5%. Both all-cause 28-day mortality and 28-day mortality related to *A. baumannii* infection were significantly lower in the group of patients that received nebulized colistin: all-cause 28-day mortality was 52.4% in Group A vs. 78.9% in Group B, p 0.034, and 28-day mortality attributed to *A. baumannii* was 40% vs. 73.5% in Group A and B, respectively, p 0.025) (Figure 1).

Group A patients receiving nebulized colistin also had a significantly reduced 7-day mortality (9.5% vs. 47.4%, $p = 0.003$), a shorter resolution of sepsis by day 7 (38.1% vs. 13.5%, $p = 0.023$) and were more likely to be without the need for vasopressors by day 7 (28.6% vs. 8.1%, $p = 0.039$). Among the two groups, there was no statistical difference in terms of complications, i.e., acute respiratory distress syndrome, coagulopathy, liver dysfunction and sepsis. Patients that were not treated with nebulized colistin had a tendency to develop more frequent septic shock, though results failed to reach statistical significance (66.7% in Group A vs. 86.8% in Group B patients, $p = 0.065$). Regarding safety considerations, colistin treatment was generally well-tolerated in both patient groups, and only in one patient in Group B, intravenous colistin was discontinued prematurely due to nephrotoxicity; the development of acute kidney injury and the need for renal replacement therapy was similar in both groups (33.3% in Group A vs. 40% in Group B, $p = 0.587$).

Table 2. Outcome and complications of the patients with bacteremic VAP from *Acinetobacter baumannii* that were treated (Group A) and not treated (Group B) with nebulized colistin.

Parameter	Group A (Treated with Nebulized Colistin) (n = 21)	Group B (Treated without Nebulized) Colistin (n = 38)	p-Value
Mortality			
Total Mortality at 28 days * (n, %)	11 (52.4%)	30 (78.9%)	0.034
Mortality at 28 days related to infection * (n, %)	6 (40%)	25 (73.5%)	0.025
Mortality at 7 days * (n, %)	2 (9.5%)	18 (47.4%)	0.003
Days on MV ∞ median (min–max)	28 (11–56)	17 (1–115))	0.053
Ventilator free days @28 d (n, %)	4 (19%)	2 (7.8%)	0.465
Median min–max	9 (4–18)	18.5 (17–20)	0.587
ICU days median (min–max)	30 (11–106)	17 (1–144)	0.007
Morbidity related to *A. baumannii* infection (n, %)	8 (38.1%)	8 (21.1%)	0.159
Free of symptoms at day 7 *	9 (45%)	10 (26.3%)	0.233
Free of vasopressors at day 7 *	6 (28.6%)	3 (8.1%)	0.039
Resolution of sepsis at day 7 *	8 (38.1%)	5 (13.5%)	0.023
Microbial cure at day 7 *	9 (45%)	10 (26.3%)	0.150
Complications related to *A. baumanii* infection (n, %)			

Table 2. *Cont.*

Parameter	Group A (Treated with Nebulized Colistin) (n = 21)	Group B (Treated without Nebulized) Colistin (n = 38)	p-Value
Sepsis	19 (90.5%)	36 (94.7%)	0.533
Shock	14 (66.7%)	33 (86.8%)	0.065
Cardiomyopathy ʃ	1 (4.8%)	3 (8.1%)	0.629
AKI ∞,ʃ	7 (33.3%)	15 (40.5%)	0.587
Coagulopathy ʃ	13 (61.9%)	29 (76.3%)	0.242
Hepatic dysfunction ʃ	11 (52.4%)	19 (50%)	0.861
ARDS ∞,ʃ	18 (85.7%)	36 (94.7%)	0.233
CRRT ∞,ʃ	2 (9.5%)	4 (10.8%)	0.877

∞ Abbreviations: ARDS, Acute Respiratory Distress Syndrome; AKI, Acute Kidney Injury; CRRT, Continuous Renal Replacement Therapy; MV, mechanical ventilation. ʃ Coagulopathy was defined as a platelet count < 150 × 10^3/μL or an INR > 1.4. Cardiomyopathy was defined as new onset left ventricular systolic dysfunction during the septic episode. Hepatic dysfunction was considered when total bilirubin was >2 mg/dL. AKI was defined by the KDIGO criteria. * Total mortality at 7 days and total mortality at 28 days were defined as death occurring between the day of a positive blood culture for *A. baumannii* up until 7 or 28 days, respectively. Mortality was attributed to *A. baumannii* if at least two intensive care physicians decided that the patient died from septic shock caused by the *A. baumannii* infection. Free of symptoms by day 7 was defined as resolution of symptoms and improvement of oxygenation by day 7 post the *A. baumannii* positive blood culture, free of vasopressors by day 7 was defined as the patient not receiving any vasopressors on the seventh day post the bacteremia. Microbiological cure by day 7 was defined as *A. baumannii* not growing in follow-up blood cultures by day 7. Resolution of sepsis was defined as a SOFA score on day 7 ≤ 24 h before the sepsis-onset SOFA plus maximum one more point.

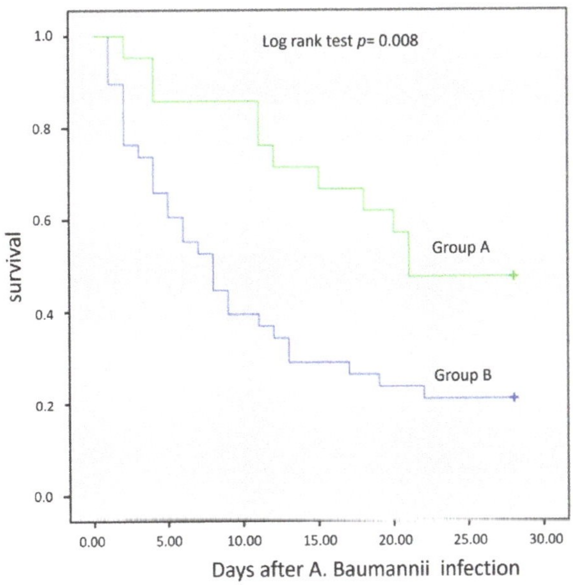

Figure 1. Probability of survival for the first 28 days post *Acinetobacter baumannii* bacteremic VAP was significantly higher for patients receiving nebulized colistin (Group A) vs. patients not receiving nebulized colistin as part of their therapeutic regime.

Finally, we performed a sensitivity analysis using a propensity score analysis. A propensity score covariate was formed and adjusted for baseline characteristics (age, disease severity and comorbidities), and a multivariate binary logistic regression analysis was performed, that included the propensity score covariate, with the PDR, nebulized colistin and high dose ampicillin/sulbactam used as covariates. The analysis showed that nebulized colistin was an independent risk factor for mortality (Supplemental Table S1).

3. Discussion

Our study of 59 critically ill patients with DTR *A. baumannii* VAP and associated bacteremia showed that the co-administration of nebulized colistin in the standard treatment regime was accompanied by a reduction in the 28-day all-cause mortality (52.4% vs. 78.9%, $p = 0.025$). It also showed that the 28-day mortality attributed to DTR *A. baumannii* and the 7-day mortality were significantly lower in the patient group treated with nebulized colistin (40% and 9.5% vs. 73.5% and 47.4%, respectively). An important finding of our study was also that those patients treated with nebulized colistin were more likely to be off vasopressors by day 7 (28.6% vs. 8.1%) and were more likely to have resolution of their sepsis by day 7 (38.1% vs. 13.5%, respectively). Interestingly, the great majority of patients (90.5% Group A, 71.1% Group B) had VAP from colistin-resistant *A. baumannii*. Also, as expected by the PK characteristics of nebulized colistin and in line with previous studies, the beneficial effects of nebulized colistin co-administration were not accompanied by an increased incidence of acute kidney injury and need for renal replacement therapy [31,32].

To the best of our knowledge, this is the first study that shows improvement in mortality in patients with bacteremic *A. baumannii* VAP receiving nebulized colistin in addition to an intravenous antibiotic regime that, in the vast majority of cases, includes colistin. Two previous studies, the study of Tumbarello et al. and the study of Korbila et al., did show a higher clinical cure rate in the nebulized group but failed to show improvement in mortality [33,34]. A possible explanation for this difference compared to our study might be the low dose of nebulized colistin used in both studies (<3 million units/day) compared to the high dose of 15 million units/day used in our study. Another important difference between our study and the two previous studies was the sensitivity pattern of *A. baumannii*. Specifically, in the studies of Tumbarello et al. and Korbila et al., *A. baumannii* strains were colistin-sensitive, while the majority of the strains in our study were colistin-resistant, and possibly, the addition of nebulized colistin in our cohort, with its associated significant increase in concentration in the epithelial lining fluid, might have made the difference in the overall efficacy of the combined regime. Finally, another factor explaining the difference in mortality outcomes might be the fact that the majority of our patients had COVID-19 pneumonia on ICU admission (90.5% Group A, 89.5% Group B). COVID-19 patients have a diffuse disease of their lung parenchyma and are more vulnerable to superinfections as a significant part of their lung parenchyma is diseased. The addition of nebulized colistin in these patients might significantly enhance the action of colistin compared to non-COVID-19 patients. Unfortunately, the number of non-COVID-19 patients with bacteremic VAP was very low (10%), and no comparison could have been made between COVID-19 and non-COVID-19 patients with bacteremic VAP.

Our study showed that patients receiving nebulized colistin also demonstrated improved 7-day outcomes, in terms of mortality as well as sepsis resolution and weaning off the vasopressors. This is in accordance with most previous studies that have showed better clinical response in patients receiving nebulized colistin [33–36]. The improvement in 7-day mortality better reflects the outcome of the infection as it is closer to the onset of the infection compared to day 28 [12].

Finally, to the best of our knowledge, our study is the largest one reporting results on the effect of nebulized colistin in colistin-resistant *A. baumannii* VAP, while most studies usually exclude patients with colistin-resistant *A. baumannii* [35]. There are only two other studies addressing this issue in PDR *A. baumannii* patients with VAP: the first is a case series study of 10 patients, who received nebulized and intravenous colistin, high-dose tigecycline and high-dose ampicillin/sulbactam, which showed a very good clinical outcome; and the second one recruited 32 patients with PDR *A. baumannii* VAP: 10 patients were treated with the above mentioned triple antibiotic combination (80% also received fosfomycin) and 22 patients received a combination of two antibiotics (details were not mentioned) [30,37].

The latter study also reported a favorable outcome with improved survival. Therefore, all three studies show a favorable outcome in patients with colistin-resistant VAP treated with nebulized colistin. As colistin resistance of *A. baumannii* strains continues to rise and new antibiotics such as cefiderocol are not yet widely available across countries (as it was also not available in our country during the study period), more studies addressing the effect of colistin nebulization in colistin-resistant *A. baumannii* VAP are urgently needed.

3.1. Limitations

Our study has limitations. First of all, as a retrospective single center study, it is subject to the inherent limitations of such studies. Our study sample is not large; however, it is similar to most studies published on the same subject [27]. Also, the majority of our cohort were COVID-19 patients, and one might argue that more data from non-COVID-19 patients are needed before we extrapolate our conclusions to a non-COVID-19 ICU population. Nevertheless, there is currently no evidence that COVID-19 patients respond differently to nebulized antibiotics. In addition, the *A. baumannii* strains of our cohort were mostly colistin-resistant, and our results might not be generalizable in different settings. Also, there was no record in patient data about which nebulizer was used in each particular patient (nebulized colistin was delivered either with vibrating mesh or jet nebulizer). Nevertheless, despite the fact that, in most cases, a jet nebulizer was used, which is known to provide nebulization in the least optimal way, [38] our study had a positive outcome for nebulized colistin. Finally, as cefiderocol was not available in our country during the study period, our results may not be generalizable in patients receiving cefiderocol as part of their therapeutic regime; nevertheless, results are conflicting regarding cefiderocol's superiority over colistin for such infections [39].

3.2. Strengths

To the best of our knowledge, this is the first study comparing patients receiving either a combination of both intravenous and high-dose nebulized colistin or intravenous colistin only. It is also the first study showing improvement in mortality outcomes. Moreover, in most previous studies, nebulized colistin was administered in critically ill patients with gram negative infections attributed to colistin-sensitive *A. baumannii*, while in our study, most patients had colistin-resistant *A. baumannii*; nearly 30% of Group A patients had PDR *A. baumannii*. The presence of a significant percentage of colistin-resistant and PDR *A. baumannii* strains reflects the current (unfortunate) reality in a significant number of ICUs in several countries globally as it has been shaped since the COVID-19 pandemic. The emergence of colistin resistance or failure in particular could be related to subtherapeutic levels of colistin in patients, underlining perhaps the importance of using nebulized colistin in VAP [40].

4. Materials and Methods

We conducted a retrospective, single center, observational study that included all patients with bacteremic VAP due to *A. baumannii* who were admitted in the two intensive care units (ICU) of Ioannina University Hospital from March 2020 to August 2023. University Hospital of Ioannina is a 900-bed academic hospital in Ioannina, Greece. Data were collected in an anonymized way, and data collection was approved by the institutional review board of the University Hospital of Ioannina (protocol number 37/2023). Due to the retrospective, observational nature of the study, the need for informed consent was waived.

4.1. Data Collection and Patient Groups

We recorded information on comorbidities, sex, age, severity of disease, including severity of disease scores on admission such as the Sequential Organ Failure Assessment (SOFA) and the Acute Physiologic Assessment and Chronic Health Evaluation (APACHE) II. We also recorded the SOFA score the previous day of sepsis onset, on the day of sepsis development and on day 7 since the development of sepsis. In addition, we recorded complications, including acute respiratory distress syndrome (ARDS), acute kidney injury (AKI), the application of continuous renal replacement therapy (CRRT), coagulopathy, sepsis, septic shock and hepatic dysfunction. We also recorded 7-day outcomes such as resolution of sepsis, discontinuation of vasopressors and microbiological cure, and long term outcomes such as 28-day mortality, association of death with the *A. baumannii* infection, antibiotic sensitivities and the type, dose and duration of antibiotics administered, ICU length of stay (LOS) and duration of mechanical ventilation. Patients that did not receive antibiotics with potential action against *A. baumannii* or received antibiotics for less than 24 h were excluded from our study. In patients with more than one episode of bacteremia from *A. baumannii*, only the first episode was recorded.

Patients were categorized into two groups: in Group A were patients with VAP and a secondary *A. baumannii* bacteremia that received nebulized colistin as part of their antibiotic treatment regime, and in Group B were patients with VAP and a secondary *A. baumannii* bacteremia that did not receive nebulized colistin as part of their antibiotic treatment regime.

4.2. Study Outcomes

The primary outcomes of the study were 28-day total mortality and 28-day mortality related to the *A. baumannii* infection. Secondary outcomes included the development of sepsis, the 7-day mortality, the 7-day resolution of sepsis, the 7-day need for vasopressors, the 7-day microbiological cure, the length of stay in the ICU and in the hospital, the duration of mechanical ventilation and the development of complications-organ failures.

4.3. Definitions

We defined ventilator-associated pneumonia (VAP) by the Centers for Disease Control and Prevention (CDC) criteria [41]. SEPSIS-3 definition was used to define sepsis and septic shock [42]. The Berlin definition was used to define ARDS [43]. For patients already diagnosed with ARDS on admission, both a worsening of the severity of the ARDS stage and a deterioration of the lung opacities were considered as new ARDS complicating the *A. baumannii* VAP [43]. We defined acute kidney injury (AKI) based on the 'Kidney Disease: Improving Global Outcomes' (KDIGO) guidelines [44]. Hepatic dysfunction was defined, as previously described, as total bilirubin greater than 2 mg/dL [12]. Coagulopathy was defined as, either a platelet count of less than 150,000/μL, or an International Normalized Ratio (INR) of more than 1.4 [45]. Death was attributed to *A. baumannii* if at least two ICU physicians concluded that the patient died from the *A. baumannii*-studied VAP. A SOFA score on day 7 from sepsis less or equal to the 24 h before sepsis SOFA score was defined as resolution of sepsis [41]. Microbiological cure was defined as a negative blood culture result for *A. baumannii* in a subsequent blood culture taken up to day 7 post the episode of bacteremia. The European Committee on Antimicrobial Susceptibility Testing (EUCAST) minimum inhibitory capacity (MIC) breakpoints were used to determine sensitivity of antibiotics [46]. MIC for colistin and tigecycline were determined by using broth microdilution, while for the rest of antibiotics the VITEK© system was used. DTR was defined as *A. baumannii* resistant to all first class antibiotics (beta-lactams, including carbapenems and fluoroquinolones [47].

4.4. Statistical Analysis

Statistical analysis was performed as previously described [12]. The Kolgomorov–Smirnov test was used to assess for normal distribution of continuous variables, and variables were presented as either median, min–max or mean ± standard deviation values as appropriate. The study groups were compared for the continuous variables either by the *t*-test or Mann–Whitney test for the continuous variables whereas, for categorical variables, χ^2 tests (p values were extracted from Fisher's exact test) were applied. The log-rank test was used to construct and compare the Kaplan–Meier survival curves of the two groups, after the proportionality of hazards was assessed graphically and confirmed for all covariates. A *p*-value of less than 0.05 was used to define statistical significance in all cases. A sensitivity analysis was performed using propensity score analysis. A propensity score was adjusted for baseline characteristics (age, severity of disease, comorbidities) and used as covariate in binary logistic regression analysis to assess for risk factors of mortality. SPSS® 29.0 was used for all statistical analyses.

5. Conclusions

The addition of high-dose nebulized colistin to the standard intravenous treatment of critically ill patients with bacteremic *A. baumannii* VAP seems to improve patients' outcome. The co-administration of high-dose nebulized colistin appears to be both beneficial and safe irrespective of the in vitro sensitivity of colistin against *A. baumannii*.

Supplementary Materials: The following supporting information can be downloaded at: https://www.mdpi.com/article/10.3390/antibiotics13020169/s1, Table S1: Propensity score-adjusted analysis on A. baumannii attributed mortality.

Author Contributions: Conceptualization, I.A.; methodology, I.A. and A.P.; software, A.P.; validation, I.A., T.M., N.L., N.K. and A.P.; formal analysis, I.A. and A.P.; investigation, I.A., T.M., N.L., N.K., E.T., G.P., K.G. and A.P.; resources, K.G., I.A., T.M., N.L., N.K. and A.P.; data curation, I.A., T.M., N.L., E.T., N.K., D.K., G.P., K.G. and A.P.; writing—original draft preparation, I.A., A.P., D.K. and G.P.; writing—review and editing, I.A., T.M., N.L., N.K., K.G., A.P., E.T., D.K., V.K. and G.P.; visualization, A.P.; supervision, I.A. and V.K. All authors have read and agreed to the published version of the manuscript.

Funding: This research received no external funding.

Institutional Review Board Statement: The study was conducted in accordance with the Declaration of Helsinki and approved by the Institutional Review Board (or Ethics Committee) of the University Hospital of Ioannina (protocol code 37/2023 and date of approval).

Informed Consent Statement: Patient consent was waived due to the retrospective and observational nature of the study.

Data Availability Statement: The datasets used and/or analyzed during the current study are available from the corresponding author on reasonable request.

Acknowledgments: Vassiliki Stamati for technical support and Maria Mpaziou for administrative support.

Conflicts of Interest: The authors declare no conflicts of interest.

References

1. Cavallo, I.; Oliva, A.; Pages, R.; Sivori, F.; Truglio, M.; Fabrizio, G.; Pasqua, M.; Pimpinelli, F.; Di Domenico, E.G. *Acinetobacter baumannii* in the critically ill: Complex infections get complicated. *Front. Microbiol.* **2023**, *14*, 1196774. [CrossRef]
2. Koulenti, D.; Vandana, K.E.; Rello, J. Current viewpoint on the epidemiology of nonfermenting Gram-negative bacterial strains. *Curr. Opin. Infect. Dis.* **2023**, *36*, 545–554. [CrossRef]
3. Scaglione, V.; Reale, M.; Davoli, C.; Mazzitelli, M.; Serapide, F.; Lionello, R.; La Gamba, V.; Fusco, P.; Bruni, A.; Procopio, D.; et al. Prevalence of Antibiotic Resistance Over Time in a Third-Level University Hospital. *Microb. Drug Resist.* **2022**, *28*, 425–435. [CrossRef]
4. Tacconelli, E.; Carrara, E.; Savoldi, A.; Harbarth, S.; Mendelson, M.; Monnet, D.L.; Pulcini, C.; Kahlmeter, G.; Kluytmans, J.; Carmeli, Y.; et al. Discovery, research, and development of new antibiotics: The WHO priority list of antibiotic-resistant bacteria and tuberculosis. *Lancet Infect. Dis.* **2018**, *18*, 318–327. [CrossRef]
5. Russo, A.; Bassetti, M.; Ceccarelli, G.; Carannante, N.; Losito, A.R.; Bartoletti, M.; Corcione, S.; Granata, G.; Santoro, A.; Giacobbe, D.R.; et al. Bloodstream infections caused by carbapenem-resistant Acinetobacter baumannii: Clinical features, therapy and outcome from a multicenter study. *J. Infect.* **2019**, *79*, 130–138. [CrossRef]
6. Du, X.; Xu, X.; Yao, J.; Deng, K.; Chen, S.; Shen, Z.; Yang, L.; Feng, G. Predictors of mortality in patients infected with carbapenem-resistant Acinetobacter baumannii: A systematic review and meta-analysis. *Am. J. Infect. Control* **2019**, *47*, 1140–1145. [CrossRef]
7. Zhou, H.; Yao, Y.; Zhu, B.; Ren, D.; Yang, Q.; Fu, Y.; Yu, Y.; Zhou, J. Risk factors for acquisition and mortality of multidrug-resistant *Acinetobacter baumannii* bacteremia: A retrospective study from a Chinese hospital. *Medicine* **2019**, *98*, e14937. [CrossRef]

8. Adukauskiene, D.; Ciginskiene, A.; Adukauskaite, A.; Koulenti, D.; Rello, J. Clinical Features and Outcomes of Monobacterial and Polybacterial Episodes of Ventilator-Associated Pneumonia Due to Multidrug-Resistant *Acinetobacter baumannii*. *Antibiotics* **2022**, *11*, 892. [CrossRef]
9. Agbaht, K.; Diaz, E.; Muñoz, E.; Lisboa, T.; Gomez, F.; Depuydt, P.O.; Blot, S.I.; Rello, J. Bacteremia in patients with ventilator-associated pneumonia is associated with increased mortality: A study comparing bacteremic vs. nonbacteremic ventilator-associated pneumonia. *Crit. Care Med.* **2007**, *35*, 2064–2070. [CrossRef]
10. Loyola-Cruz, M.Á.; Gonzalez-Avila, L.U.; Martínez-Trejo, A.; Saldaña-Padilla, A.; Hernández-Cortez, C.; Bello-López, J.M.; Castro-Escarpulli, G. ESKAPE and Beyond: The Burden of Coinfections in the COVID-19 Pandemic. *Pathogens* **2023**, *12*, 743. [CrossRef]
11. Segala, F.V.; Pafundi, P.C.; Masciocchi, C.; Fiori, B.; Taddei, E.; Antenucci, L.; De Angelis, G.; Guerriero, S.; Pastorino, R.; Damiani, A.; et al. Incidence of bloodstream infections due to multidrug-resistant pathogens in ordinary wards and intensive care units before and during the COVID-19 pandemic: A real-life, retrospective observational study. *Infection* **2023**, *51*, 1061–1069. [CrossRef]
12. Andrianopoulos, I.; Maniatopoulou, T.; Lagos, N.; Kazakos, N.; Papathanasiou, A.; Papathanakos, G.; Koulenti, D.; Kittas, C.; Koulouras, V. *Acinetobacter baumannii* Bloodstream Infections in the COVID-19 Era: A Comparative Analysis between COVID-19 and Non-COVID-19 Critically Ill Patients. *Microorganisms* **2023**, *11*, 1811. [CrossRef]
13. Finazzi, S.; Luci, G.; Olivieri, C.; Langer, M.; Mandelli, G.; Corona, A.; Viaggi, B.; Di Paolo, A. Tissue Penetration of Antimicrobials in Intensive Care Unit Patients: A Systematic Review—Part I. *Antibiotics* **2022**, *11*, 1164. [CrossRef]
14. Viaggi, B.; Cangialosi, A.; Langer, M.; Olivieri, C.; Gori, A.; Corona, A.; Finazzi, S.; Di Paolo, A. Tissue Penetration of Antimicrobials in Intensive Care Unit Patients: A Systematic Review—Part II. *Antibiotics* **2022**, *11*, 1193. [CrossRef]
15. Bassetti, M.; Luyt, C.-E.; Nicolau, D.P.; Pugin, J. Characteristics of an ideal nebulized antibiotic for the treatment of pneumonia in the intubated patient. *Ann. Intensive Care* **2016**, *6*, 35. [CrossRef]
16. Leng, B.; Yan, G.; Wang, C.; Shen, C.; Zhang, W.; Wang, W. Dose optimisation based on pharmacokinetic/pharmacodynamic target of tigecycline. *J. Glob. Antimicrob. Resist.* **2021**, *25*, 315–322. [CrossRef]
17. De Pascale, G.; Lisi, L.; Ciotti, G.M.P.; Vallecoccia, M.S.; Cutuli, S.L.; Cascarano, L.; Gelormini, C.; Bello, G.; Montini, L.; Carelli, S.; et al. Pharmacokinetics of high-dose tigecycline in critically ill patients with severe infections. *Ann. Intensive Care* **2020**, *10*, 94. [CrossRef]
18. Mantzana, P.; Protonotariou, E.; Kassomenaki, A.; Meletis, G.; Tychala, A.; Keskilidou, E.; Arhonti, M.; Katsanou, C.; Daviti, A.; Vasilaki, O.; et al. In Vitro Synergistic Activity of Antimicrobial Combinations against Carbapenem- and Colistin-Resistant *Acinetobacter baumannii* and *Klebsiella pneumoniae*. *Antibiotics* **2023**, *12*, 93. [CrossRef]
19. Wang, S.-H.; Yang, K.-Y.; Sheu, C.-C.; Lin, Y.-C.; Chan, M.-C.; Feng, J.-Y.; Chen, C.-M.; Chen, C.-Y.; Zheng, Z.-R.; Chou, Y.-C.; et al. The prevalence, presentation and outcome of colistin susceptible-only Acinetobacter Baumannii-associated pneumonia in intensive care unit: A multicenter observational study. *Sci. Rep.* **2023**, *13*, 140. [CrossRef]
20. Yang, K.-Y.; Peng, C.-K.; Sheu, C.-C.; Lin, Y.-C.; Chan, M.-C.; Wang, S.-H.; Chen, C.-M.; Chen, C.-Y.; Zheng, Z.-R.; Feng, J.-Y.; et al. Clinical effectiveness of tigecycline in combination therapy against nosocomial pneumonia caused by CR-GNB in intensive care units: A retrospective multi-centre observational study. *J. Intensive Care* **2023**, *11*, 1. [CrossRef]
21. Heil, E.L.; Claeys, K.C.; Kline, E.G.; Rogers, T.M.; Squires, K.M.; Iovleva, A.; Doi, Y.; Banoub, M.; Noval, M.M.; Luethy, P.M.; et al. Early initiation of three-drug combinations for the treatment of carbapenem-resistant *A. baumannii* among COVID-19 patients. *J. Antimicrob. Chemother.* **2023**, *78*, 1034–1040. [CrossRef]
22. Dickstein, Y.; Lellouche, J.; Ben Dalak Amar, M.; Schwartz, D.; Nutman, A.; Daitch, V.; Yahav, D.; Leibovici, L.; Skiada, A.; Antoniadou, A.; et al. Treatment Outcomes of Colistin- and Carbapenem-resistant *Acinetobacter baumannii* Infections: An Exploratory Subgroup Analysis of a Randomized Clinical Trial. *Clin. Infect. Dis.* **2019**, *69*, 769–776. [CrossRef]
23. Kaye, K.S.; Marchaim, D.; Thamlikitkul, V.; Carmeli, Y.; Chiu, C.-H.; Daikos, G.; Dhar, S.; Durante-Mangoni, E.; Gikas, A.; Kotanidou, A.; et al. Colistin Monotherapy versus Combination Therapy for Carbapenem-Resistant Organisms. *NEJM Evid.* **2023**, *2*, EVIDoa2200131. [CrossRef]
24. De Pascale, G.; Pintaudi, G.; Lisi, L.; De Maio, F.; Cutuli, S.L.; Tanzarella, E.S.; Carelli, S.; Lombardi, G.; Cesarano, M.; Gennenzi, V.; et al. Use of High-Dose Nebulized Colistimethate in Patients with Colistin-Only Susceptible *Acinetobacter baumannii* VAP: Clinical, Pharmacokinetic and Microbiome Features. *Antibiotics* **2023**, *12*, 125. [CrossRef]
25. Heffernan, A.J.; Sime, F.B.; Lipman, J.; Dhanani, J.; Andrews, K.; Ellwood, D.; Grimwood, K.; Roberts, J.A. Intrapulmonary pharmacokinetics of antibiotics used to treat nosocomial pneumonia caused by Gram-negative bacilli: A systematic review. *Int. J. Antimicrob. Agents* **2019**, *53*, 234–245. [CrossRef]
26. Zhang, X.; Cui, X.; Jiang, M.; Huang, S.; Yang, M. Nebulized colistin as the adjunctive treatment for ventilator-associated pneumonia: A systematic review and meta-analysis. *J. Crit. Care* **2023**, *77*, 154315. [CrossRef]
27. Karaiskos, I.; Gkoufa, A.; Polyzou, E.; Schinas, G.; Athanassa, Z.; Akinosoglou, K. High-Dose Nebulized Colistin Methanesulfonate and the Role in Hospital-Acquired Pneumonia Caused by Gram-Negative Bacteria with Difficult-to-Treat Resistance: A Review. *Microorganisms* **2023**, *11*, 1459. [CrossRef]
28. Papathanakos, G.; Andrianopoulos, I.; Papathanasiou, A.; Koulenti, D.; Gartzonika, K.; Koulouras, V. Pandrug-resistant *Acinetobacter baumannii* treatment: Still a debatable topic with no definite solutions. *J. Antimicrob. Chemother.* **2020**, *75*, 3081. [CrossRef]

29. Karakonstantis, S.; Kritsotakis, E.I.; Gikas, A. Pandrug-resistant Gram-negative bacteria: A systematic review of current epidemiology, prognosis and treatment options. *J. Antimicrob. Chemother.* **2020**, *75*, 271–282. [CrossRef]
30. Assimakopoulos, S.F.; Karamouzos, V.; Lefkaditi, A.; Sklavou, C.; Kolonitsiou, F.; Christofidou, M.; Fligou, F.; Gogos, C.; Marangos, M. Triple combination therapy with high-dose ampicillin/sulbactam, high-dose tigecycline and colistin in the treatment of ventilator-associated pneumonia caused by pan-drug resistant *Acinetobacter baumannii*: A case series study. *Infez. Med.* **2019**, *27*, 11–16.
31. Min, K.L.; Son, E.S.; Kim, J.S.; Kim, S.H.; Jung, S.M.; Chang, M.J. Risk factors of colistin safety according to administration routes: Intravenous and aerosolized colistin. *PLoS ONE* **2018**, *13*, e0207588. [CrossRef] [PubMed]
32. Park, S.Y.; Park, M.S.; Chung, C.R.; Kim, J.S.; Park, S.J.; Lee, H.B. Clinical Effectiveness and Nephrotoxicity of Aerosolized Colistin Treatment in Multidrug-Resistant Gram-Negative Pneumonia. *Korean J. Crit. Care Med.* **2016**, *31*, 208–220. [CrossRef]
33. Tumbarello, M.; De Pascale, G.; Trecarichi, E.M.; De Martino, S.; Bello, G.; Maviglia, R.; Spanu, T.; Antonelli, M. Effect of Aerosolized Colistin as Adjunctive Treatment on the Outcomes of Microbiologically Documented Ventilator-Associated Pneumonia Caused by Colistin-Only Susceptible Gram-Negative Bacteria. *Chest* **2013**, *144*, 1768–1775. [CrossRef]
34. Korbila, I.P.; Michalopoulos, A.; Rafailidis, P.I.; Nikita, D.; Samonis, G.; Falagas, M.E. Inhaled colistin as adjunctive therapy to intravenous colistin for the treatment of microbiologically documented ventilator-associated pneumonia: A comparative cohort study. *Clin. Microbiol. Infect.* **2010**, *16*, 1230–1236. [CrossRef]
35. Feng, J.-Y.; Huang, J.-R.; Lee, C.-C.; Tseng, Y.-H.; Pan, S.-W.; Chen, Y.-M.; Yang, K.-Y. Role of nebulized colistin as a substitutive strategy against nosocomial pneumonia caused by CR-GNB in intensive care units: A retrospective cohort study. *Ann. Intensive Care* **2023**, *13*, 1. [CrossRef]
36. Kim, Y.K.; Lee, J.H.; Lee, H.-K.; Chung, B.C.; Yu, S.J.; Lee, H.-Y.; Park, J.-H.; Kim, S.; Kim, H.-K.; Kiem, S.; et al. Efficacy of nebulized colistin-based therapy without concurrent intravenous colistin for ventilator-associated pneumonia caused by carbapenem-resistant Acinetobacter baumannii. *J. Thorac. Dis.* **2017**, *9*, 555–567. [CrossRef]
37. Casarotta, E.; Bottari, E.; Vannicola, S.; Giorgetti, R.; Domizi, R.; Carsetti, A.; Damiani, E.; Scorcella, C.; Gabbanelli, V.; Pantanetti, S.; et al. Antibiotic Treatment of *Acinetobacter baumannii* Superinfection in Patients With SARS-CoV-2 Infection Admitted to Intensive Care Unit: An Observational Retrospective Study. *Front. Med.* **2022**, *9*, 910031. [CrossRef]
38. The European Investigators Network for Nebulized Antibiotics in Ventilator-associated Pneumonia; Rouby, J.J.; Sole-Lleonart, C.; Rello, J. Ventilator-associated pneumonia caused by multidrug-resistant Gram-negative bacteria: Understanding nebulization of aminoglycosides and colistin. *Intensive Care Med.* **2020**, *46*, 766–770. [CrossRef]
39. Mazzitelli, M.; Gregori, D.; Sasset, L.; Trevenzoli, M.; Scaglione, V.; Lo Menzo, S.; Marinello, S.; Mengato, D.; Venturini, F.; Tiberio, I.; et al. Cefiderocol-Based versus Colistin-Based Regimens for Severe Carbapenem-Resistant *Acinetobacter baumannii* Infections: A Propensity Score-Weighted, Retrospective Cohort Study during the First Two Years of the COVID-19 Pandemic. *Microorganisms* **2023**, *11*, 984. [CrossRef]
40. Marano, V.; Marascio, N.; Pavia, G.; Lamberti, A.G.; Quirino, A.; Musarella, R.; Casalinuovo, F.; Mazzitelli, M.; Trecarichi, E.M.; Torti, C.; et al. Identification of pmrB mutations as putative mechanism for colistin resistance in *A. baumannii* strains isolated after in vivo colistin exposure. *Microb. Pathog.* **2020**, *142*, 104058. [CrossRef]
41. Magill, S.S.; Klompas, M.; Balk, R.; Burns, S.M.; Deutschman, C.S.; Diekema, D.; Fridkin, S.; Greene, L.; Guh, A.; Gutterman, D.; et al. Developing a New, National Approach to Surveillance for Ventilator-Associated Events. *Crit. Care Med.* **2013**, *41*, 2467–2475. [CrossRef]
42. Singer, M.; Deutschman, C.S.; Seymour, C.W.; Shankar-Hari, M.; Annane, D.; Bauer, M.; Bellomo, R.; Bernard, G.R.; Chiche, J.-D.; Coopersmith, C.M.; et al. The Third International Consensus Definitions for Sepsis and Septic Shock (Sepsis-3). *JAMA* **2016**, *315*, 801. [CrossRef] [PubMed]
43. Ranieri, V.I.T.O.; Rubenfeld, G.D.; Thompson, B.T.; Ferguson, N.D.; Caldwell, E.; Fan, E.; Camporota, L. Acute Respiratory Distress Syndrome: The Berlin Definition. *JAMA* **2012**, *307*, 2526–2533. [CrossRef] [PubMed]
44. Kellum, J.A.; Lameire, N.; KDIGO AKI Guideline Work Group. Diagnosis, evaluation, and management of acute kidney injury: A KDIGO summary (Part 1). *Crit. Care* **2013**, *17*, 204. [CrossRef]
45. Perez Ruiz de Garibay, A.; Kortgen, A.; Leonhardt, J.; Zipprich, A.; Bauer, M. Critical care hepatology: Definitions, incidence, prognosis and role of liver failure in critically ill patients. *Crit. Care* **2022**, *26*, 289. [CrossRef]
46. The European Committee on Antimicrobial Susceptibility Testing. Breakpoint Tables for Interpretation of MICs and Zone Diameters. Version 13.1. 2023. Available online: http://www.eucast.org (accessed on 10 December 2023).
47. Kadri, S.S.; Adjemian, J.; Lai, Y.L.; Spaulding, A.B.; Ricotta, E.; Prevots, D.R.; Palmore, T.N.; Rhee, C.; Klompas, M.; Dekker, J.P.; et al. Difficult-to-Treat Resistance in Gram-negative Bacteremia at 173 US Hospitals: Retrospective Cohort Analysis of Prevalence, Predictors, and Outcome of Resistance to All First-line Agents. *Clin. Infect. Dis.* **2018**, *67*, 1803–1814. [CrossRef]

antibiotics

MDPI

Article

Evaluation of Fosfomycin-Sulbactam Combination Therapy against Carbapenem-Resistant *Acinetobacter baumannii* Isolates in a Hollow-Fibre Infection Model

Sazlyna Mohd Sazlly Lim [1,2], Aaron Heffernan [1,3], Saiyuri Naicker [1], Steven Wallis [1], Jason A. Roberts [1,4,5] and Fekade Bruck Sime [1,*]

1 UQ Centre for Clinical Research, Faculty of Medicine, University of Queensland, Brisbane, QLD 4029, Australia
2 Department of Medicine, Faculty of Medicine and Health Sciences, Universiti Putra Malaysia, Serdang 43400, Malaysia
3 School of Medicine, Griffith University, Southport, QLD 4222, Australia
4 Departments of Pharmacy and Intensive Care Medicine, Royal Brisbane and Women's Hospital, Brisbane, QLD 4029, Australia
5 Division of Anaesthesiology Critical Care Emergency and Pain Medicine, Nîmes University Hospital, University of Montpellier, 30900 Nîmes, France
* Correspondence: f.sime@uq.edu.au; Tel.: +61-7-3346-5023

Abstract: Static concentration in vitro studies have demonstrated that fosfomycin- or sulbactam-based combinations may be efficacious against carbapenem-resistant *Acinetobacter baumannii* (CRAB). In the present study, we aimed to evaluate the bacterial killing and resistance suppression potential of fosfomycin-sulbactam combination therapies against CRAB isolates in a dynamic infection model. We simulated clinically relevant dosing regimens of fosfomycin (8 g every 8 h, 1 h infusion) and sulbactam (12 g continuous infusion or 4 g every 8 h, 4 h infusion) alone and in combination for 7 days in a hollow-fibre infection model (HFIM) against three clinical isolates of CRAB. The simulated pharmacokinetic profiles in the HFIM were based on fosfomycin and sulbactam data from critically ill patients. Fosfomycin monotherapy resulted in limited bacterial killing. Sulbactam monotherapies resulted in ~ 3 to 4 $\log_{10}$ kill within the first 8 to 32 h followed by regrowth of up to 8 to 10 $\log_{10}$ CFU/mL. A combination of fosfomycin and continuous infusion of sulbactam led to a ~2 to 4 $\log_{10}$ reduction in bacterial burden within the first 24 h, which was sustained throughout the duration of the experiments. A combination of fosfomycin and extended infusion of sulbactam produced a ~4 $\log_{10}$ reduction in colony count within 24 h. This study demonstrated that fosfomycin in combination with sulbactam is a promising option for the treatment of MDR *A. baumannii*. Further studies are needed to further assess the potential clinical utility of this combination.

Keywords: fosfomycin; sulbactam; combination; hollow-fibre infection model; *Acinetobacter baumannii*; carbapenem-resistant; pharmacodynamic

1. Introduction

Acinetobacter baumannii is a difficult pathogen to effectively treat as it has the ability to protect itself in the face of antibiotic therapy via various mechanisms of resistance [1]. Carbapenems have previously been the agent of choice to counteract species producing extended spectrum beta-lactamase (ESBL), including *A. baumannii*. However, carbapenems are losing viability for the treatment of *A. baumannii* infections as the prevalence of carbapenem-resistant *A. baumannii* (CRAB) increases [2]. Furthermore, many of the newer antibiotics under development or available on the pharmaceutical market for drug-resistant Gram-negative bacilli are active against KPC-producers or carbapenem-resistant *P. aeruginosa* but not against CRAB [3]. Except for cefiderocol, other newer agents lack activity against *A. baumannii* [4].

Considering the limited antibiotics that are available for the treatment of CRAB infections and the paucity of new agents that are active against CRAB strains, antibiotic combination therapy appears to be an appealing treatment alternative. Most antibiotic combination studies for the treatment of multidrug-resistant (MDR) *A. baumannii* have looked at polymyxin-based combinations [4]. However, the use of polymyxin is often limited by its narrow therapeutic window and nephrotoxicity [5], making polymyxin-based combinations less than ideal. An ideal antibiotic combination would be one that is active against the bacteria of interest, able to suppress the emergence of resistance and has a better safety profile.

Fosfomycin- or sulbactam-based combinations may be considered as prospective treatment substitutes against CRAB. High-dose (≥4 g/day) sulbactam-based combinations have been shown to result in better clinical improvement and clinical cure compared to colistin-based combinations in the treatment of MDR *A. baumannii* [6]. In terms of the safety profile of high-dose sulbactam, studies have demonstrated that sulbactam doses of 9–12 g/day can be used safely in a clinical setting [7,8].

Numerous fosfomycin- and sulbactam-based combinations have been studied in vitro at fixed concentrations [9–11]. Synergism has been observed in various fosfomycin- and sulbactam-based combinations against MDR *A. baumannii* with synergy rates ranging from 30 to 80% [12–15]. In one of our earlier reports, we explored the fosfomycin and sulbactam combination in checkerboard and static time-kill studies [16]. Two-drug combination checkerboard assays demonstrated that the fosfomycin-sulbactam combination was synergistic against 37/50 CRAB (74%) isolates. This finding was further supported by static time-kill experiments, which demonstrated that the combination was synergistic at fosfomycin and sulbactam concentrations of 128 mg/L, respectively, for two out of two isolates tested.

Despite compelling data seen in static in vitro infection models, the role of combination therapy such as fosfomycin and sulbactam in a clinical setting remains ambiguous, as the dynamicity of the in vitro models and in vivo conditions clearly differs. Therefore, in the present study, we set out to evaluate the combination of fosfomycin and sulbactam against CRAB isolates in a dynamic infection model. We assessed the bacterial killing activity and suppression of resistance of the fosfomycin and sulbactam combination at clinically relevant plasma exposures.

2. Materials and Methods

2.1. Bacterial Isolates

Fifty isolates were obtained from The University of Queensland Centre of Clinical Research (UQCCR). These isolates were chosen from samples previously analysed by Zowawi et al. [17]. These 50 isolates were used to screen for synergistic activity against various antibiotic combinations, including the fosfomycin and sulbactam combination, in a previous study [16]. Out of the fifty isolates, three isolates were chosen for the hollow-fibre infection model (HFIM) experiments.

The isolates were stored in cation-adjusted Mueller-Hinton II broth (CA-MH) (Becton, Dickinson and Company, Sparks, MD, USA) containing 20% glycerol in a freezer at −80 °C. Before each experiment, fresh isolates were streaked out and grown on Mueller-Hinton agar (Becton, Dickinson and Company, Sparks, MD, USA) plates, incubated at 37 °C for 18 h and used for the preparation of the inoculum. Carbapenem resistance was determined by Zowawi et al. [17] using disk diffusion susceptibility testing for imipenem (10 µg) and meropenem (10 µg) following the European Committee on Antimicrobial Susceptibility Testing (EUCAST) methodology.

2.2. Antimicrobial Agents

Meropenem (Tokyo Chemical Industry Co. Ltd., Tokyo, Japan), sulbactam (Acros Organics, Lot: A0405260, Shanghai, China) and fosfomycin (Wako Pure Chemical Industries, Ltd., Osaka, Japan) were obtained from their respective manufacturers. Stock solutions of these antibiotics were prepared in sterile Milli-Q water, filter-sterilised with a 0.22 μm polyvinylidene difluoride (PVDF) syringe filter, aliquoted and stored at −80 °C until required. Before each susceptibility test, an aliquot of the drug was thawed and diluted to the desired concentrations with cation-adjusted Mueller-Hinton II (CA-MH) broth. Broth or agar containing fosfomycin was supplemented with 25 mg/L of glucose-6-phosphate (Sigma-Aldrich, St. Louis, MO, USA).

2.3. In Vitro Susceptibility

The MICs of meropenem and sulbactam against the three CRAB isolates were determined by the broth microdilution method, in quadruplicate, following the recommendations of CLSI, as described in the CLSI M100 approved standard [18]. The mode MIC was reported for each isolate. Fosfomycin susceptibility testing was performed by agar dilution [18]. *Klebsiella pneumoniae* (ATCC 700603), *Pseudomonas aeruginosa* (ATCC 27853) and *Enterococcus faecalis* (ATCC 29212) strains were used as quality control strains for sulbactam, fosfomycin and meropenem, respectively. The fosfomycin and sulbactam MIC of the isolates from all the experimental arms were also reassessed at the end of the experiment using the same methods.

2.4. Hollow-Fibre Infection Model

The HFIM was put together using FiberCell Systems cellulosic cartridges (C3008) in a 37 °C incubator as detailed previously [19]. In the experiments investigating the fosfomycin-sulbactam combination, a supplementing compartment was introduced to simulate the differential clearance of the two antibiotics as described by Blaser [20]. The CA-MH broth used in all experiments was supplemented with 25 mg/L glucose-6-phosphate. All experiments were conducted over seven days. One HFIM experiment was conducted for each dosing regimen and isolate combination. For each isolate, an overnight bacterial suspension was prepared in CA-MH broth at 36 °C and subsequently diluted to achieve an initial inoculum of approximately 6.5 $\log_{10}$ CFU/mL.

Sulbactam exhibits primarily time-dependent bacterial killing [21] and has been shown to be 93% stable for 24 h at 37 °C [22]. Therefore, a continuous or extended infusion of sulbactam could maximise pharmacodynamic target attainment. For fosfomycin, reports on the pharmacodynamic index associated with bacterial killing are mixed. Some studies have reported that fosfomycin exhibits time-dependent activity against *Staphylococcus aureus* and *Escherichia coli* [23,24]. Other more recent studies on fosfomycin demonstrated concentration-dependent bacterial killing against *E. coli* [25,26]. In the present study, considering concentration-dependent activity and the ratio of the area under the concentration-time curve to the MIC (AUC/MIC) as the driver of efficacy for fosfomycin, an intermittent infusion at a high dose was chosen to maximise this index. The following treatment regimens were simulated in the HFIM experiments: (1) fosfomycin 8 g every 8 h, given as a 1 h infusion (intermittent infusion); (2) sulbactam 12 g daily, given as a continuous infusion (preceded by a 4 g loading dose, given as a 1 h infusion); (3) sulbactam 4 g every 8 h, given as a 4 h infusion (extended infusion). These fosfomycin and sulbactam regimens were simulated alone and in combination in the HFIM. A loading dose of sulbactam (4 g, 1 h infusion) was given prior to the continuous infusion of sulbactam to rapidly achieve steady-state concentration. A no-treatment growth control was also included.

The simulated pharmacokinetic profiles in the HFIM were based on the pharmacokinetic models of fosfomycin and sulbactam developed for critically ill patients [27,28]. To ensure the simulated profiles are representative of the central location of the target profile, a normal patient body weight of 70 kg and a creatinine clearance of 100 mL/min were used in the covariate models for the estimation of primary PK parameters. For fosfomycin, negligible protein binding [29], a 4 h elimination half-life [27] and peak concentration of 300 mg/L were used to simulate free drug exposures. For sulbactam, 38% protein binding [30], a 1.5 h elimination half-life [28] and peak concentration of 60 mg/L were used to simulate free-drug exposures. The simulations were deemed acceptable if the best-fit peak concentrations and elimination half-lives were both within 20% of the target values [31]. A one-compartment model was fitted to the observed sulbactam and fosfomycin concentration-time profiles using the Pmetrics package (v1.52, Laboratory of Applied Pharmacokinetics and Bioinformatics, Los Angeles, CA, USA) for R (v.3.6.2, R Core Team, Vienna, Austria).

The clearance of each drug was simulated by controlling the flow rates of the CA-MH broth flowing through the inner lumens of the hollow-fibre cartridge. Peristaltic pumps (Masterflex L/S; Cole-Parmer, Vernon Hills, IL, USA) were used to provide CA-MH broth flow through the central reservoir. A Duet Pump (FiberCell Systems Inc., New Market, MD, USA) was used to provide continuous circulation between the central and peripheral compartments to ensure continuous media and drug distribution.

2.5. *Quantification of Viable Bacterial Populations*

Bacterial quantification was performed with periodic sampling at 0, 2, 4, 6, 8, 24, 32, 48, 72, 96, 120, 144 and 168 h from the cartridge's extracapillary space sampling port. Samples were washed twice in phosphate-buffered saline to minimise antibiotic carry-over. A 100 μL aliquot of an appropriately diluted bacterial suspension was manually plated onto CA-MH agar. The limit of quantification was 2-$\log_{10}$ CFU/mL.

2.6. *Fosfomycin and Sulbactam Assays for Pharmacokinetics*

Samples were taken from the central compartment sampling port at 1, 2, 4, 8, 24, 25, 28, 32, 48, 49, 52, 56, 72, 96, 120, 144 and 168 h for pharmacokinetic analysis to ascertain whether the intended pharmacokinetic profiles of the antibiotics were achieved in the HFIM. These samples were immediately stored at −80 °C until bioanalysis.

Concentrations of fosfomycin and sulbactam in CA-MH broth were measured over a calibration range of 1 to 1000 mg/L by a validated ultra-high performance liquid chromatography using a tandem mass spectrometry (UHPLC-MS/MS) method on a Nexera2 UHPLC system coupled with a 8030+ triple quadrupole mass spectrometer (Shimadzu, Kyoto, Japan). Test samples were assayed alongside plasma calibrators and quality controls and met the batch acceptance criteria (US FDA 2018).

For fosfomycin, the CA-MH broth (20 μL) was spiked with internal standard (ethylphosphonic acid), vortex mixed and centrifuged. The supernatant was injected into the UHPLC-MS/MS instrument. The stationary phase was a SeQuant zic-HILIC 2.1 × 50 mm (5.0 μm) analytical column protected by a 20 mm SeQuant zic-HILIC guard cartridge (Merck, Darmstadt, Germany) operated at room temperature. Mobile phase A was 1 mM ammonium acetate (pH 4.5), and mobile phase B was 100% acetonitrile with 0.2% formic acid (v/v). The method was isocratic (25% mobile phase B). Fosfomycin was monitored by negative mode electrospray at m/z of 137.0→79.0. Ethylphosphonic acid was monitored in negative mode at 109.05→79.0. The lower limit of quantitation (LLOQ) was 1 mg/L. Inter-batch precision was 4.4, 6.2 and 7.0% at 5, 50 and 5000 mg/L, respectively. Inter-batch accuracy was 5.5, 2.1 and −4.4% at 5, 50 and 5000 mg/L, respectively.

For sulbactam, the CA-MH broth sample (10 μL) was spiked with internal standard (tazobactam) and protein was precipitated with acetonitrile. An aliquot of 1 μL of the supernatant was injected into the UHPLC-MS/MS. The stationary phase was Shim-pack XR-ODS III 2 × 50 mm (1.6 μm) analytical column (Shimadzu, Kyoto, Japan) preceded by a C18 UHPLC analytical guard column (Phenomenex, Torrence, CA, USA). Mobile phase A was 0.1% formic acid (v/v), and mobile phase B was 100% acetonitrile with 0.1% formic acid (v/v). The mobile phase was delivered as a gradient from 10% B to 75% B and back again at a flow 0.35 mL/min, producing a backpressure of approximately 8500 psi. Ionisation was performed by electrospray. Sulbactam was monitored in negative mode at 232.05→140.05. The internal standard tazobactam was monitored in negative mode at 299.00→138.15. The assay method was linear from 0.78 to 156 mg/L (LLOQ of 0.78 mg/L). Inter-assay precision was 6.7, 1.5 and 13.6% and accuracy was 1.7, −7.0 and −10.8% at 2.7, 18 and 144 mg/L sulbactam in CA-MH broth, respectively.

3. Results

3.1. In Vitro Susceptibility

The fosfomycin MIC for isolates #75, #98 and #102 was 128, 256 and 128 mg/L, respectively. The sulbactam MIC was 64 mg/L for all three isolates. These isolates were confirmed to be carbapenem-resistant with a meropenem MIC ranging between 32 to 64 mg/L.

3.2. Hollow-Fibre Infection Model

The fosfomycin and sulbactam pharmacokinetic profiles were reasonably well simulated in the HFIM. Typical profiles for fosfomycin and sulbactam regimens are shown in Figure 1. The predicted versus observed pharmacokinetic profiles of 8 g fosfomycin every 8 h (1 h infusion), 4 g sulbactam loading dose (1 h infusion) and 12 g continuous infusion every 24 h and 4 g sulbactam every 8 h (4 h infusion) showed reasonable agreement with r^2 values of 0.89, 0.92 and 0.92, respectively. Changes in viable counts are shown in Figure 2. The fosfomycin monotherapy resulted in limited bacterial killing against isolates #75 and #102 (stasis within the first 4 to 8 h, followed by growth up to ~10 $\log_{10}$ CFU/mL). Against isolate #98, fosfomycin monotherapy led to a 3 $\log_{10}$ reduction in bacterial burden by 8 h, followed by regrowth. The sulbactam continuous infusion in monotherapy resulted in ~3 to 4 $\log_{10}$ kill within the first 8 to 32 h followed by regrowth up to 8 to 10 $\log_{10}$ CFU/mL. The sulbactam extended infusion regimen yielded a ~3 to 4 $\log_{10}$ kill by 8 to 32 h after the start of infusion, with subsequent regrowth of up to 10 $\log_{10}$ CFU/mL.

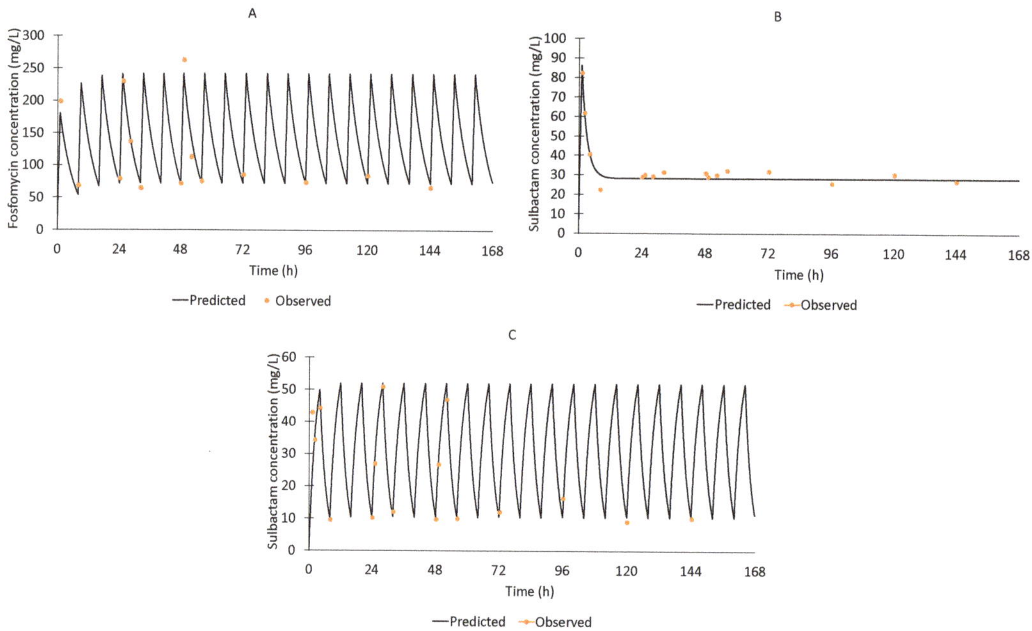

Figure 1. Typical simulated pharmacokinetic profile of (**A**) 8 g fosfomycin every 8 h (1 h infusion) (r^2 = 0.89); (**B**) 4 g sulbactam loading dose (1 h infusion) and 12 g continuous infusion every 24 h (r^2 = 0.92); (**C**) 4 g sulbactam every 8 h (4 h infusion) (r^2 = 0.92).

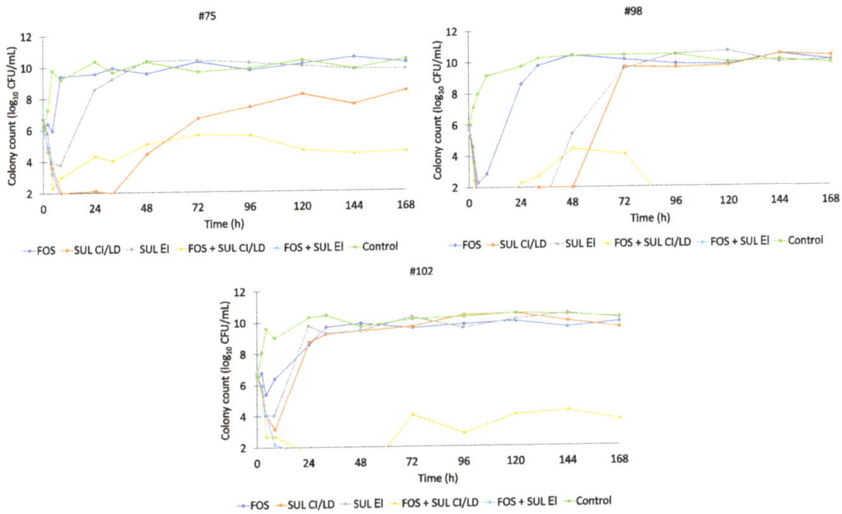

Figure 2. Time-kill curves for isolates #75, #98 and #102. FOS, fosfomycin; SUL, sulbactam; EI, extended infusion; CI, continuous infusion; LD, loading dose.

The combination of fosfomycin and continuous infusion of sulbactam led to a ~2 to 4 $\log_{10}$ reduction in the bacterial burden within the first 24 h, which was sustained throughout the duration of the experiments for isolates #75 and #102. The combination of fosfomycin and extended infusion of sulbactam, on the other hand, produced a ~4 $\log_{10}$ reduction in colony count within 24 h followed by suppression of regrowth.

3.3. *In Vitro Susceptibility Post-Drug Exposure*

Table 1 summarises the fosfomycin and sulbactam MICs post exposure. Exposure to fosfomycin monotherapy led to an increase in the fosfomycin MIC by 3 to 4 fold; whereas, the sulbactam MIC remained the same. A 2 fold increase in the sulbactam MIC was observed after exposure to sulbactam monotherapies; whereas, the fosfomycin MIC remained the same despite exposure to sulbactam. Exposure to the combination of fosfomycin and continuous infusion of sulbactam did not affect the subsequent sulbactam MIC of the isolates. We did, however, observe a 2-fold rise in the fosfomycin MIC. The post-exposure in vitro susceptibility testing for the fosfomycin and extended infusion of sulbactam combination arms were not performed as there was no colony growth on the CA-MH agar.

Table 1. Fosfomycin and sulbactam minimum inhibitory concentration post-antibiotic exposure.

Isolates	Pre-Exposure MIC (mg/L)		Antibiotic Regimen	Post-Exposure MIC (mg/L)	
	Fosfomycin	Sulbactam		Fosfomycin	Sulbactam
#75	128	64	Fosfomycin monotherapy	2048	32
			Sulbactam CI monotherapy	128	64
			Sulbactam EI monotherapy	128	256
			Fosfomycin + Sulbactam CI	512	32
			Fosfomycin + Sulbactam EI	-	-
#98	256	64	Fosfomycin monotherapy	2048	64
			Sulbactam CI monotherapy	256	128
			Sulbactam EI monotherapy	128	128
			Fosfomycin + Sulbactam CI	-	-
			Fosfomycin + Sulbactam EI	-	-
#102	256	64	Fosfomycin monotherapy	1024	32
			Sulbactam CI monotherapy	128	256
			Sulbactam EI monotherapy	128	256
			Fosfomycin + Sulbactam CI	512	32
			Fosfomycin + Sulbactam EI	-	-

MIC, minimum inhibitory concentration; CI, continuous infusion; EI, extended infusion.

4. Discussion

To the best of our knowledge, this is the first study to explore the in vitro activity of fosfomycin-sulbactam combinations against CRAB in a dynamic infection model. The most intriguing finding in this study is the extent of bactericidal activity observed in the fosfomycin and sulbactam extended infusion arms against the tested isolates, despite the inherent resistance of *A. baumannii* to fosfomycin [32]. Similarly, a HFIM study by Lenhard et al. [33] demonstrated that high-dose ampicillin/sulbactam (sulbactam total daily dose 12 g/day) in combination with meropenem and polymyxin B resulted in rapid clearance of a carbapenem- and polymyxin-resistant *A. baumannii* isolates. These findings support the role of sulbactam in the treatment of MDR *A. baumannii*.

Several studies have also demonstrated benefits of fosfomycin-based combination therapy in patients with MDR *A. baumannii* infection. Sirijatuphat et al. [34] observed better microbiological eradication within 72 h of treatment (90.7% vs. 58.1%, $p = 0.001$) and a trend toward better clinical outcomes and mortality rates in the colistin-fosfomycin combination group compared with the colistin monotherapy group. Furthermore, a prospective, observational, multicentre study demonstrated that fosfomycin-containing combinations were associated with 30-day survival (hazard ratio 0.04, CI 95% 0.01–0.13, $p = 0.001$) in patients with severe pneumonia due to MDR *A. baumannii* [35]. Intravenous fosfomycin has also been shown to demonstrate a favourable safety profile, with only mild adverse effects, such as mild hypokalaemia due to high sodium load with administration of fosfomycin disodium, which did not require discontinuation of therapy [36,37]. Similarly, several studies have shown that sulbactam doses of 9–12 g/day can be used safely in a clinical setting [7,8].

As our understanding of the molecular mechanism driving sulbactam's activity against *A. baumannii* is still not complete, we are not able to fully explain the mechanism of fosfomycin-sulbactam synergy. However, the synergy is likely due to the inhibition of bacterial cell wall synthesis at two different stages in the biosynthesis pathway. Fosfomycin binds to phosphoenolpyruvate transferase thereby inhibiting mucopeptide synthesis, which is an early step in the synthesis peptidoglycan polymer that makes the cell wall [38]. It therefore acts at an earlier stage of bacterial cell wall production than most other antibiotics that inhibit cell wall synthesis [38]. Sulbactam is a class A beta-lactamase inhibitor with intrinsic whole-cell activity against certain bacterial species, including *A. baumannii* [39]. Sulbactam binds to penicillin-binding proteins of *Acinetobacter* spp., subsequently affecting a later stage of bacterial cell wall biosynthesis involving peptidoglycan cross-linkage. It is hypothesised that this activity accounts for the bactericidal effects of sulbactam on *A. baumannii* [40].

Another notable finding is the superiority of the sulbactam extended infusion regimen compared to the sulbactam continuous infusion regimen in terms of bacterial killing and suppression of resistance against CRAB isolates seen in this study. Neither the sulbactam extended infusion nor the continuous infusion regimens achieved concentrations that exceeded the sulbactam MIC of 64 mg/L (Figure 1). Nonetheless, the sulbactam extended infusion regimen performed better, probably due to the relatively higher concentration it achieved (C_{max} ~50 mg/L), compared to the continuous infusion regimen (steady-state concentration 25–30 mg/L). Lenhard et al. [33] demonstrated that an ampicillin/sulbactam dosing regimen of 8/4 g every 8 h, when simulated in the HFIM, was able to maintain a percentage of time above the MIC of 74% for sulbactam against *A. baumannii* isolates with sulbactam MIC 16 mg/L, which likely explained the drastic killing observed during ampicillin/sulbactam monotherapy in their study. However, considering the high sulbactam MIC of the CRAB isolates used in the present study, the sulbactam extended infusion was not able to maintain a decent percentage of time above the MIC, as the simulated dosing regimen did not achieve concentrations that exceeded the sulbactam MIC of 64 mg/L (Figure 1). Nonetheless, the combination containing the sulbactam extended infusion was still able to exhibit bacterial killing and suppression of regrowth.

It was expected that the CRAB isolates tested would develop further resistance to fosfomycin and sulbactam following exposure to fosfomycin and sulbactam monotherapy [41]. Interestingly, we observed an increase in fosfomycin MIC, but not sulbactam MIC, following exposure to the combination therapy, in which sulbactam was given as a continuous infusion. As combination regimens containing 4 g sulbactam every 8 h (extended infusion) and 12 g continuous infusion exhibited differing abilities to prevent further fosfomycin resistance, we hypothesised that this observation was likely due to the difference in the sulbactam concentration at steady state between the two regimens (sulbactam extended infusion vs. continuous infusion). It appears that sulbactam is able to prevent the further development of fosfomycin resistance in these isolates, in a concentration-dependent manner. The exact mechanism by which this phenomenon occurs should be studied further.

In this study, we used CRAB isolates that have high MICs (fosfomycin MIC 128–256 mg/L and sulbactam MIC 64 mg/L) to mirror difficult-to-treat, worst-case scenarios. The sulbactam MIC of the isolates used in our HFIM experiments were much higher compared with the MICs reported in other studies [42,43]. For fosfomycin, studies have shown that 70–80% of CRAB isolates have fosfomycin MICs of 128–256 mg/L [13,44]. This finding is supported by our own study, where we found that 66% of our CRAB isolates had similar fosfomycin MICs of 128–256 mg/L [45].

The dosing regimens simulated were chosen on the basis of our in silico study [46], which demonstrated the highest probability of target attainment of 2 $\log_{10}$ kill (71.6%) when doses of fosfomycin 24 g/day and sulbactam 12 g/day were used in combination. Attainment of 2 $\log_{10}$ kill is important, as discussed by Drusano et. al. [47], as it is this magnitude of bacterial burden reduction that is associated with near-maximal granulocyte-mediated bacterial kill. Total daily doses were limited to those used in a clinical setting as the goal was to identify optimal regimens for quick assimilation into clinical practice. Additionally, these HFIM experiments represent a conservative estimate of bactericidal activity as this infection model lacks an immune response and therefore does not account for the effect of the native immune system.

Nonetheless, this study is not without its limitations. Firstly, our study was performed against three clinical CRAB isolates, and therefore the findings might not be applicable to all MDR *A. baumannii* isolates. However, as observed in our previous checkerboard study, the fosfomycin-sulbactam combination is synergistic in a good fraction of CRAB isolates (~70%) [16]; therefore, the current findings are likely indicative of microbiological outcomes in a large proportion of CRAB isolates. Future studies should be carried out using MDR *A. baumannii* isolates with varying fosfomycin and sulbactam MICs to confirm these findings.

Secondly, whilst we acknowledge that the quantification of emergent resistant strains using agar plates containing the antibiotic at concentrations several times higher than the MIC is a common way to explore effects on resistant subpopulations, this was not feasible in our study as the isolates used were already resistant with high fosfomycin and sulbactam MICs. However, we have assessed the MIC of fosfomycin and sulbactam post-treatment to identify any shift from the baseline. Thirdly, we were unable to explain the difference in the response to the treatment arms between the three isolates. It has been discussed previously [4] that there is limited understanding to date as to what type of strains are more susceptible to combination therapy. Indeed, this may be a significant reason for the lack of reproducibility of results from in vitro studies.

Additionally, we only tested dosing regimens requiring 24 g/day of fosfomycin and 12 g/day of sulbactam, albeit these regimens are clinically accepted. Further studies, in which the doses of fosfomycin and sulbactam are titrated to establish a dose-response curve, are required to fully inform rational dosing of this combination. The time-kill profile of the fosfomycin and sulbactam combination presented in this study may also vary at different sites of infection, particularly considering the differences in site-specific pharmacokinetics in humans. Further studies in controlled clinical trials are needed to evaluate the use of this combination for the treatment of CRAB infections.

5. Conclusions

In conclusion, this study demonstrated that fosfomycin in combination with sulbactam is a promising option in our fight against MDR *A. baumannii*. We found that fosfomycin in combination with extended infusion of sulbactam was the most active of all the regimens evaluated against three CRAB isolates in a seven-day HFIM. The bacterial clearance observed in the HFIM suggests that clinicians may be able to prevent the emergence of resistance during therapy through dosing regimen modifications and the strategic selection of combination regimens.

Author Contributions: Conceptualization, S.M.S.L., F.B.S. and J.A.R.; methodology, S.M.S.L., F.B.S., S.N. and S.W.; formal analysis, S.M.S.L., A.H., F.B.S. and S.W.; resources, F.B.S. and J.A.R.; data curation, S.M.S.L.; writing—original draft preparation, S.M.S.L.; writing—review & editing, A.H., S.N., S.W., F.B.S. and J.A.R.; visualization, S.M.S.L.; supervision, S.W., F.B.S. and J.A.R.; project administration, S.M.S.L. and F.B.S.; funding acquisition, J.A.R. All authors have read and agreed to the published version of the manuscript.

Funding: This work was supported by the Australian National Health and Medical Research Council for a Centre of Research Excellence fund (APP1099452). J.A.R is funded in part by a Practitioner Fellowship (APP1117065) from the National Health and Medical Research Council of Australia (NHMRC). S.M.S.L. acknowledges funding from the University of Queensland Research Training Scholarship. F.B.S. acknowledges funding from the NHMRC Investigator Grant APP1197866.

Institutional Review Board Statement: Not applicable.

Informed Consent Statement: Not applicable.

Data Availability Statement: Not applicable.

Conflicts of Interest: The authors declare no conflict of interest.

References

1. Lee, C.-R.; Lee, J.H.; Park, M.; Park, K.S.; Bae, I.K.; Kim, Y.B.; Cha, C.-J.; Jeong, B.C.; Lee, S.H. Biology of *Acinetobacter baumannii*: Pathogenesis, antibiotic resistance mechanisms, and prospective treatment options. *Front. Cell. Infect. Microbiol.* **2017**, *7*, 55. [CrossRef] [PubMed]
2. Peleg, A.Y.; Seifert, H.; Paterson, D.L. *Acinetobacter baumannii*: Emergence of a successful pathogen. *Clin. Microbiol. Rev.* **2008**, *21*, 538–582. [CrossRef] [PubMed]
3. Isler, B.; Doi, Y.; Bonomo, R.A.; Paterson, D.L. New treatment options against carbapenem-resistant *Acinetobacter baumannii* infections. *Antimicrob. Agents Chemother.* **2019**, *63*, e01110-18. [CrossRef]
4. Mohd Sazlly Lim, S.; Sime, F.B.; Roberts, J.A. Multidrug-resistant *Acinetobacter baumannii* infections: Current evidence on treatment options and the role of pharmacokinetics/pharmacodynamics in dose optimisation. *Int. J. Antimicrob. Agents* **2019**, *53*, 726–745. [CrossRef] [PubMed]
5. Sorlí, L.; Luque, S.; Grau, S.; Berenguer, N.; Segura, C.; Montero, M.M.; Álvarez-Lerma, F.; Knobel, H.; Benito, N.; Horcajada, J.P. Trough colistin plasma level is an independent risk factor for nephrotoxicity: A prospective observational cohort study. *BMC Infect. Dis.* **2013**, *13*, 380. [CrossRef] [PubMed]
6. Liu, J.; Zhu, F.; Feng, B.; Zhang, Z.; Liu, L.; Wang, G. Comparative efficacy and safety of combination therapy with high-dose sulbactam or colistin with additional antibacterial agents for multiple drug-resistant and extensively drug-resistant *Acinetobacter baumannii* infections: A systematic review and network meta-analysis. *J. Glob. Antimicrob. Resist.* **2021**, *24*, 136–147. [CrossRef]
7. Betrosian, A.P.; Frantzeskaki, F.; Xanthaki, A.; Douzinas, E.E. Efficacy and safety of high-dose ampicillin/sulbactam vs. colistin as monotherapy for the treatment of multidrug resistant *Acinetobacter baumannii* ventilator-associated pneumonia. *J. Infect.* **2008**, *56*, 432–436. [CrossRef]
8. Betrosian, A.P.; Frantzeskaki, F.; Xanthaki, A.; Georgiadis, G. High-dose ampicillin-sulbactam as an alternative treatment of late-onset VAP from multidrug-resistant *Acinetobacter baumannii*. *Scand. J. Infect. Dis.* **2007**, *39*, 38–43. [CrossRef]
9. Falagas, M.E.; Kastoris, A.C.; Karageorgopoulos, D.E.; Rafailidis, P.I. Fosfomycin for the treatment of infections caused by multidrug-resistant non-fermenting Gram-negative bacilli: A systematic review of microbiological, animal and clinical studies. *Int. J. Antimicrob. Agents* **2009**, *34*, 111–120. [CrossRef]
10. Chen, H.; Liu, Q.; Chen, Z.; Li, C. Efficacy of sulbactam for the treatment of *Acinetobacter baumannii* complex infection: A systematic review and meta-analysis. *J. Infect. Chemother. Off. J. Jpn. Soc. Chemother.* **2017**, *23*, 278–285. [CrossRef]
11. Flamm, R.K.; Rhomberg, P.R.; Lindley, J.M.; Sweeney, K.; Ellis-Grosse, E.J.; Shortridge, D. Evaluation of the Bactericidal Activity of Fosfomycin in Combination with Selected Antimicrobial Comparison Agents Tested against Gram-Negative Bacterial Strains by Using Time-Kill Curves. *Antimicrob. Agents Chemother.* **2019**, *63*, e02549-18. [CrossRef] [PubMed]

12. Santimaleeworagun, W.; Wongpoowarak, P.; Chayakul, P.; Pattharachayakul, S.; Tansakul, P.; Garey, K.W. In vitro activity of colistin or sulbactam in combination with fosfomycin or imipenem against clinical isolates of carbapenem-resistant *Acinetobacter baumannii* producing OXA-23 carbapenemases. *Southeast Asian J. Trop. Med. Public Health* **2011**, *42*, 890. [PubMed]
13. Zhu, W.; Wang, Y.; Cao, W.; Cao, S.; Zhang, J. In vitro evaluation of antimicrobial combinations against imipenem-resistant *Acinetobacter baumannii* of different MICs. *J. Infect. Public Health* **2018**, *11*, 856–860. [CrossRef] [PubMed]
14. Wei, W.; Yang, H.; Liu, Y.; Ye, Y.; Li, J. In vitro synergy of colistin combinations against extensively drug-resistant *Acinetobacter baumannii* producing OXA-23 carbapenemase. *J. Chemother.* **2016**, *28*, 159–163. [CrossRef]
15. Deveci, A.; Coban, A.Y.; Acicbe, O.; Tanyel, E.; Yaman, G.; Durupinar, B. In vitro effects of sulbactam combinations with different antibiotic groups against clinical *Acinetobacter baumannii* isolates. *J. Chemother.* **2012**, *24*, 247–252. [CrossRef]
16. Mohd Sazlly Lim, S.; Naicker, S.; Ayfan, A.K.; Zowawi, H.; Roberts, J.A.; Sime, F.B. Non-polymyxin-based combinations as potential alternatives in treatment against carbapenem-resistant *Acinetobacter baumannii* infections. *Int. J. Antimicrob. Agents* **2020**, *56*, 106115. [CrossRef]
17. Zowawi, H.M.; Sartor, A.L.; Sidjabat, H.E.; Balkhy, H.H.; Walsh, T.R.; Al Johani, S.M.; AlJindan, R.Y.; Alfaresi, M.; Ibrahim, E.; Al-Jardani, A. Molecular Epidemiology of carbapenem resistant *Acinetobacter baumannii* in the Gulf Cooperation Council States. Dominance of OXA-23-type producers. *J. Clin. Microbiol.* **2015**, *53*, 896–903. [CrossRef]
18. Clinical and Laboratory Standards Institute. Performance Standards for Antimicrobial Susceptibility Testing 30th ed. Available online: http://em100.edaptivedocs.net/dashboard.aspx (accessed on 8 June 2020).
19. Cadwell, J. The Hollow Fiber Infection Model: Principles and Practice. *Adv. Antibiot. Antibodies* **2015**, *1*, 2.
20. Blaser, J. In-vitro model for simultaneous simulation of the serum kinetics of two drugs with different half-lives. *J. Antimicrob. Chemother.* **1985**, *15*, 125–130. [CrossRef]
21. Rodríguez-Hernández, M.-J.; Cuberos, L.; Pichardo, C.; Caballero, F.J.; Moreno, I.; Jiménez-Mejías, M.E.; García-Curiel, A.; Pachón, J. Sulbactam efficacy in experimental models caused by susceptible and intermediate *Acinetobacter baumannii* strains. *J. Antimicrob. Chemother.* **2001**, *47*, 479–482. [CrossRef]
22. Wildfeuer, A.; Räder, K. Stability of β-lactamase inhibitors and β-lactam antibiotics in parenteral dosage forms and in body fluids and tissue homogenates: A comparative study of sulbactam, clavulanic acid, ampicillin and amoxycillin. *Int. J. Antimicrob. Agents* **1996**, *6*, S31–S34. [CrossRef]
23. Pfausler, B.; Spiss, H.; Dittrich, P.; Zeitlinger, M.; Schmutzhard, E.; Joukhadar, C. Concentrations of fosfomycin in the cerebrospinal fluid of neurointensive care patients with ventriculostomy-associated ventriculitis. *J. Antimicrob. Chemother.* **2004**, *53*, 848–852. [CrossRef] [PubMed]
24. VanScoy, B.D.; McCauley, J.; Ellis-Grosse, E.J.; Okusanya, O.O.; Bhavnani, S.M.; Forrest, A.; Ambrose, P.G. Exploration of the Pharmacokinetic-Pharmacodynamic Relationships for Fosfomycin Efficacy Using an In Vitro Infection Model. *Antimicrob. Agents Chemother.* **2015**, *59*, 7170–7177. [CrossRef] [PubMed]
25. Zykov, I.N.; Samuelsen, Ø.; Jakobsen, L.; Småbrekke, L.; Andersson, D.I.; Sundsfjord, A.; Frimodt-Møller, N. Pharmacokinetics and Pharmacodynamics of Fosfomycin and Its Activity against Extended-Spectrum-β-Lactamase-, Plasmid-Mediated AmpC-, and Carbapenemase-Producing Escherichia coli in a Murine Urinary Tract Infection Model. *Antimicrob. Agents Chemother.* **2018**, *62*, e02560-17. [CrossRef]
26. Docobo-Pérez, F.; Drusano, G.L.; Johnson, A.; Goodwin, J.; Whalley, S.; Ramos-Martín, V.; Ballestero-Tellez, M.; Rodriguez-Martinez, J.M.; Conejo, M.C.; van Guilder, M.; et al. Pharmacodynamics of Fosfomycin: Insights into Clinical Use for Antimicrobial Resistance. *Antimicrob. Agents Chemother.* **2015**, *59*, 5602. [CrossRef] [PubMed]
27. Joukhadar, C.; Klein, N.; Dittrich, P.; Zeitlinger, M.; Geppert, A.; Skhirtladze, K.; Frossard, M.; Heinz, G.; Müller, M. Target site penetration of fosfomycin in critically ill patients. *J. Antimicrob. Chemother.* **2003**, *51*, 1247–1252. [CrossRef]
28. Jaruratanasirikul, S.; Nitchot, W.; Wongpoowarak, W.; Samaeng, M.; Nawakitrangsan, M. Population pharmacokinetics and Monte Carlo simulations of sulbactam to optimize dosage regimens in patients with ventilator-associated pneumonia caused by *Acinetobacter baumannii. Eur. J. Pharm. Sci.* **2019**, *136*, 104940. [CrossRef]
29. Zeitlinger, M.; Sauermann, R.; Traunmüller, F.; Georgopoulos, A.; Müller, M.; Joukhadar, C. Impact of plasma protein binding on antimicrobial activity using time–killing curves. *J. Antimicrob. Chemother.* **2004**, *54*, 876–880. [CrossRef]
30. Rolinson, G.; Sutherland, R. The binding of antibiotics to serum proteins. *Br. J. Pharmacol. Chemother.* **1965**, *25*, 638–650. [CrossRef]
31. Abodakpi, H.; Chang, K.-T.; Gao, S.; Sánchez-Díaz, A.M.; Cantón, R.; Tam, V.H. Optimal piperacillin-tazobactam dosing strategies against extended-spectrum-β-lactamase-producing Enterobacteriaceae. *Antimicrob. Agents Chemother.* **2019**, *63*, e01906-18. [CrossRef]
32. Gil-Marqués, M.L.; Moreno-Martínez, P.; Costas, C.; Pachón, J.; Blázquez, J.; McConnell, M.J. Peptidoglycan recycling contributes to intrinsic resistance to fosfomycin in *Acinetobacter baumannii. J. Antimicrob. Chemother.* **2018**, *73*, 2960–2968. [CrossRef] [PubMed]
33. Lenhard, J.R.; Smith, N.M.; Bulman, Z.P.; Tao, X.; Thamlikitkul, V.; Shin, B.S.; Nation, R.L.; Li, J.; Bulitta, J.B.; Tsuji, B.T. High-dose ampicillin-sulbactam combinations combat polymyxin-resistant *Acinetobacter baumannii* in a hollow-fiber infection model. *Antimicrob. Agents Chemother.* **2017**, *61*, e01268-16. [CrossRef]
34. Sirijatuphat, R.; Thamlikitkul, V. Preliminary study of colistin versus colistin plus fosfomycin for treatment of carbapenem-resistant *Acinetobacter baumannii* infections. *Antimicrob. Agents Chemother.* **2014**, *58*, 5598–5601. [CrossRef]

35. Russo, A.; Bassetti, M.; Bellelli, V.; Bianchi, L.; Cattaneo, F.M.; Mazzocchetti, S.; Paciacconi, E.; Cottini, F.; Schiattarella, A.; Tufaro, G. Efficacy of a Fosfomycin-Containing Regimen for Treatment of Severe Pneumonia Caused by Multidrug-Resistant *Acinetobacter baumannii*: A Prospective, Observational Study. *Infect. Dis. Ther.* **2020**, *10*, 187–200. [CrossRef]
36. Grabein, B.; Graninger, W.; Baño, J.R.; Dinh, A.; Liesenfeld, D. Intravenous fosfomycin—Back to the future. Systematic review and meta-analysis of the clinical literature. *Clin. Microbiol. Infect.* **2017**, *23*, 363–372. [CrossRef]
37. Iarikov, D.; Wassel, R.; Farley, J.; Nambiar, S. Adverse events associated with fosfomycin use: Review of the literature and analyses of the FDA adverse event reporting system database. *Infect. Dis. Ther.* **2015**, *4*, 433–458. [CrossRef]
38. Kahan, F.M.; Kahan, J.S.; Cassidy, P.J.; Kropp, H. The mechanism of action of fosfomycin (phosphonomycin). *Ann. N. Y. Acad. Sci.* **1974**, *235*, 364–386. [CrossRef]
39. Noguchi, J.; Gill, M. Sulbactam: A beta-lactamase inhibitor. *Clin. Pharm.* **1988**, *7*, 37–51.
40. Papp-Wallace, K.M.; Senkfor, B.; Gatta, J.; Chai, W.; Taracila, M.A.; Shanmugasundaram, V.; Han, S.; Zaniewski, R.P.; Lacey, B.M.; Tomaras, A.P. Early insights into the interactions of different β-lactam antibiotics and β-lactamase inhibitors against soluble forms of *Acinetobacter baumannii* PBP1a and Acinetobacter sp. PBP3. *Antimicrob. Agents Chemother.* **2012**, *56*, 5687–5692. [CrossRef]
41. Cantón, R.; Morosini, M.-I. Emergence and spread of antibiotic resistance following exposure to antibiotics. *FEMS Microbiol. Rev.* **2011**, *35*, 977–991. [CrossRef]
42. Oliveira, M.S.d.; Costa, S.F.; Pedri, E.d.; van der Heijden, I.; Levin, A.S.S. The minimal inhibitory concentration for sulbactam was not associated with the outcome of infections caused by carbapenem-resistant *Acinetobacter* sp. treated with ampicillin/sulbactam. *Clinics* **2013**, *68*, 569–573. [CrossRef]
43. Chuang, Y.-C.; Sheng, W.-H.; Lauderdale, T.-L.; Li, S.-Y.; Wang, J.-T.; Chen, Y.-C.; Chang, S.-C. Molecular epidemiology, antimicrobial susceptibility and carbapenemase resistance determinants among *Acinetobacter baumannii* clinical isolates in Taiwan. *J. Microbiol. Immunol. Infect.* **2014**, *47*, 324–332. [CrossRef] [PubMed]
44. Leite, G.C.; Oliveira, M.S.; Perdigão-Neto, L.V.; Rocha, C.K.D.; Guimarães, T.; Rizek, C.; Levin, A.S.; Costa, S.F. Antimicrobial Combinations against Pan-Resistant *Acinetobacter baumannii* Isolates with Different Resistance Mechanisms. *PLoS ONE* **2016**, *11*, e0151270. [CrossRef]
45. Mohd Sazlly Lim, S.; Heffernan, A.J.; Roberts, J.A.; Sime, F.B. Pharmacodynamic Analysis of Meropenem and Fosfomycin Combination Against Carbapenem-Resistant *Acinetobacter baumannii* in Patients with Normal Renal Clearance: Can It Be a Treatment Option? *Microb. Drug Resist.* **2020**, *27*, 546–552. [CrossRef]
46. Lim, S.M.S.; Heffernan, A.; Roberts, J.; Sime, F. Semi-mechanistic PK/PD modelling of fosfomycin and sulbactam combination against carbapenem-resistant *Acinetobacter baumannii. Antimicrob. Agents Chemother.* **2021**, *65*, 5. [CrossRef]
47. Louie, A.; Brown, D.; Baluya, D.; Rodriquez, J.; Robbins, N.; Kurhanewicz, S.; Fikes, S.; Liu, W.; Drusano, G.L. Cirz R: Interaction of Drug- and Granulocyte-Mediated Killing of Pseudomonas aeruginosa in a Murine Pneumonia Model. *J. Infect. Dis.* **2014**, *210*, 1319–1324.

antibiotics

MDPI

Review

Current Therapeutic Approaches for Multidrug-Resistant and Extensively Drug-Resistant *Acinetobacter baumannii* Infections

Petros Rafailidis [1], Periklis Panagopoulos [1], Christos Koutserimpas [2] and George Samonis [3,4,*]

[1] Second University Department of Internal Medicine, University General Hospital of Alexandroupolis, 681 00 Alexandroupolis, Greece; prafaili@med.duth.gr (P.R.); ppanago@med.duth.gr (P.P.)
[2] Department of Orthopaedics and Traumatology, "251" Hellenic Air Force General Hospital of Athens, 115 25 Athens, Greece; chrisku91@hotmail.com
[3] Department of Oncology, Metropolitan Hospital, 185 47 Athens, Greece
[4] Department of Medicine, University of Crete, 715 00 Heraklion, Greece
* Correspondence: samonis@med.uoc.gr; Tel.: +30-6944437255

Abstract: The treatment of *Acinetobacter baumannii* infections remains a challenge for physicians worldwide in the 21st century. The bacterium possesses a multitude of mechanisms to escape the human immune system. The consequences of *A. baumannii* infections on morbidity and mortality, as well on financial resources, remain dire. Furthermore, *A. baumannii* superinfections have also occurred during the COVID-19 pandemic. While prevention is important, the antibiotic armamentarium remains the most essential factor for the treatment of these infections. The main problem is the notorious resistance profile (including resistance to carbapenems and colistin) that this bacterium exhibits. While newer beta lactam/beta-lactamase inhibitors have entered clinical practice, with excellent results against various infections due to Enterobacteriaceae, their contribution against *A. baumannii* infections is almost absent. Hence, we have to resort to at least one of the following, sulbactam, polymyxins E or B, tigecycline or aminoglycosides, against multidrug-resistant (MDR) and extensively drug-resistant (XDR) *A. baumannii* infections. Furthermore, the notable addition of cefiderocol in the fight against *A. baumannii* infections represents a useful addition. We present herein the existing information from the last decade regarding therapeutic advances against MDR/XDR *A. baumannii* infections.

Keywords: *A. baumannii*; antimicrobial resistance; sulbactam; colistin; polymyxin B; tigecycline; cefiderocol

1. Introduction

Acinetobacter baumannii (*A. baumannii*) received its name in 1954 (almost 40 years after its initial description by Beijerink). It gradually became evident that this immotile pathogen is associated mainly with nosocomial infections and less frequently with community-acquired ones [1]. It is notorious for its antimicrobial resistance against a variety of antibiotic classes, such as beta-lactams, quinolones, aminoglycosides and tetracyclines/glycylcyclines. It possesses a multitude of defence mechanisms: the production of beta-lactamases, especially of serine oxacillinases (Amber class D OXA type) and metallo-beta-lactamases; changes in outer membrane proteins; multidrug efflux pumps; and alterations in the affinity or expression of the Penicillin Binding Proteins. Efflux pumps and the modification of binding sites are also involved in resistance towards quinolones, while efflux- and aminoglycoside-modifying enzymes and 16S rRNA methylation have been employed by the bacterium against aminoglycosides [2]. It does not, therefore, come as a surprise that *A. baumannii* has unfortunately established itself as one of the most significant pathogens of nosocomial infections, with a special predilection for critically ill patients [3]. The Infectious Diseases Society of America has included *A. baumannii* in the ESKAPE group of bacteria. *A. baumannii* is associated with various types of infections ranging from relatively mild, such as urinary tract infections (UTIs), skin and soft tissue infections and tracheitis, to severe, such as Ventilator-associated pneumonia (VAP), bacteraemia (BSI) [associated with Central venous catheter presence or complicating various infections such as VAP and UTIs] and complicated Intra-abdominal infections [4,5]. The main problems are the severe infections as presented in a study by Pogue et al., mainly VAP and bacteraemia due to isolates with resistance to carbapenems [2,3].

Various definitions of the resistant profile of Gram-negative bacteria, including *A. baumannii*, have been formulated in order to provide harmony and better understanding of the various clinical trials that address the epidemiology and the clinical effectiveness of antimicrobial agents against these bacteria. The most renowned definitions are those by Magiorakos et al. [6] and by Kadri et al. [7]. Magiorakos et al. defined, for epidemiological reasons, multidrug-resistant (MDR) bacteria as acquired non-susceptibility to at least one agent in three or more antimicrobial categories, extensively drug-resistant (XDR) bacteria as non-susceptibility to at least one agent in all but two or fewer antimicrobial categories (i.e., bacterial isolates remain susceptible to only one or two categories) and Pandrug-resistant (PDR) bacteria as non-susceptibility to all agents in all antimicrobial categories. On the other hand, Kadri et al. proposed a more clinical definition of difficult-to-treat resistance (DTR) as resistance to all beta lactams (including carbapenems) and fluoroquinolones. It is interesting to see that, from Kadri's original cohort, the bacterium that topped the ranking of DTR was *A. baumannii* with 18.3% DTR in the USA. In other words, for 18.3 percent of *A. baumannii* there was no beta lactam and no quinolone available for treatment. Indeed, mortality was higher for patients with DTR isolates. To provide a better picture of the resistance problem of *A. baumannii* in this study, 29.4% were carbapenem-resistant, 55.4% were resistant to extended-spectrum cephalosporinases and 49.5% were resistant to fluoroquinolones. The situation in Europe and in other parts of the world is no better. While in Northern Europe carbapenem-resistant *A. baumannii* comprises from less than 1% to 10% of the population, in Southern and Eastern Europe this escalates to a staggering ≥50%. A significant *A. baumannii* global burden is evident on other continents with MDR *A. baumannii* prevalence ranging from 64.6% in Eastern Asia to 100% in Central and Latin America with a pool overall mortality rate of 42.6% [8].

One, therefore, understands that to address *A. baumannii* infections is of paramount importance. Guidelines regarding its treatment have been provided from the IDSA [9] and the European Society for Clinical Microbiology and Infection [10]. However, the guidelines do not differentiate data from the last decade from previous data. In our opinion, a focus on the latest data, as in our review, is relevant in regard to current antimicrobial resistance.

2. Methods

We sought to identify and analyse data from clinical studies regarding therapeutic options for the treatment of MDR and also XDR Acinetobacter infections during the last decade, 2014–2023. This timeframe also includes the approximately 3 years of the burden of the COVID-19 pandemic, especially the treatment of ICU patients. The Pubmed and Scopus databases were searched. Only clinical trials and articles written in English language regarding the treatment of MDR and XDR *A. baumannii* infections in adult patients (≥18 years) were included. This manuscript does not mean to be an exhaustive review of the literature in general but rather to provide data from the last decade regarding the most relevant studies. Case reports and case series with less than 10 patients were not included.

3. Results

We identified a variety of clinical trials from the last decade regarding a variety of antimicrobial agents that are effective against *A. baumannii* infections (Tables 1–5). Among them are first-line agents as described in the definition of DTR, such as beta lactam antibiotics (including beta lactam/beta-lactamase inhibitor combinations and carbapenems), and second-line agents, such as polymyxins, tetracyclines/glycylcyclines, aminoglycosides and trimethoprim/sulfamethoxazole.

Beta lactams and beta-lactamase inhibitors that have been used in the treatment of MDR, XDR and PDR *A. baumannii* infections are sulbactam, ampicillin/sulbactam and durlobactam/sulbactam, imipenem-cilastatin and meropenem. Antipseudomonal penicillins and cephalosporins such as piperacillin/tazobactam, ceftazidime and cefepime have also been referred to in the literature as possessing in vitro activity but clinical data are scarce. Meropenem and imipenem/cilastatin possess in vitro activity and are also clinically effective, while ertapenem does not. A siderophore cephalosporin, cefiderocol, has FDA approval for treatment of *A. baumannii* infections.

We present the data existing in the literature during the last decade regarding the use of these specific agents in the treatment of MDR/XDR *A. baumannii* infections and our personal perspective on their use.

3.1. Beta Lactam/Beta-Lactamase Inhibitor Combinations

3.1.1. Sulbactam

Sulbactam, a beta-lactamase inhibitor with in vitro activity against *A. baumannii*, is not usually available as a solitary pharmaceutical agent [11,12]. It is included in various fixed combinations with ampicillin (which does not possess activity against *A. baumannii*) or cefoperazone. Nevertheless, sulbactam is unique as a beta-lactamase inhibitor as it possesses intrinsic antibacterial activity against *A. baumannii* due to its ability to bind Penicillin Binding Proteins 1 and 3. Its clinical effectiveness has been corroborated in various clinical trials. Older studies (non-randomised) have investigated the clinical effectiveness of sulbactam in the treatment of *A. baumannii* infections with a variety of clinical results [13–20]. However, contemporary data by Betrosian et al. [20] have introduced the concept of high-dose ampicillin/sulbactam dosing with the administration of 9 gr every 8 h (i.e., 9 gr of sulbactam daily). Lately, sulbactam has also been combined with another beta-lactamase inhibitor, durlobactam. This pairing of two beta-lactamase inhibitors appears unusual. However, it is based on the fact that durlobactam restores the activity of sulbactam against *A. baumannii*. This is achieved by durlobactam through the inhibition of various class D beta-lactamases (OXA-23, OXA-24 and OXA-58).

A number of clinical trials have been performed mainly investigating the outcome of VAP due to the MDR *A. baumannii* complex being treated with sulbactam, ampicillin/sulbactam, cefoperazone/sulbactam or sulbactam/durlobactam and various comparator antibiotics. We summarise these eleven studies in Table 1. We identified seven non-randomised studies (five retrospective [21–25] and two prospective [26,27]) and four randomised trials [28–31]. A total of 928 patients were included in these clinical trials. Sulbactam was administered as an ampicillin/sulbactam-containing regimen versus a comparator antibiotic regimen that either did not contain sulbactam or did contain sulbactam with a different comparator antibiotic. The daily dose of sulbactam in these studies ranged from 4 g/day to 8 g/day. As stated above [20], higher doses have been provided in the past that reached 9 g/day and some have also proposed 12 g/day [32]. Nevertheless, the 4–8 g of sulbactam that was administered in the presented trials showed similar results to the comparator medication regarding clinical cure and/or improvement in five trials, better results in four trials and worse results in one trial, while no data were reported in one trial. Regarding mortality, five out of eleven studies showed lower mortality (four with statistical significance and one with a trend) in the sulbactam arm and three studies reported no difference in mortality, while two of the trials did not report data. In total, 58 out of 448 patients that received sulbactam died (mortality of 12.94%), while 162 out of 482 patients in the comparator arm died (mortality 33.6%). Thus, absolute mortality was approximately 20 percent lower in the sulbactam arm than in the comparators. The data from the seven studies that reported on microbiological eradication did not show any statistical difference when comparing sulbactam-containing regimens with comparators.

Table 1. Summary of comparative studies examining efficacy of sulbactam in patients with *A. baumannii* infections.

Study (Year)	No. of Evaluable	Infection	Antibacterial Regimens No. Pts				Cure and/or Improvement (%)			Bacteriological Eradication (%)		
			Comparator 1		Comparator 2		Comparator 1	Comparator 2	*p*-Value	Comparator 1	Comparator 2	*p*-Value
[21]	65	Pneumonia	A/S	22	A/S + V	43	63.6	65.1	0.906	89.5	81.3	0.69
							M:36.4	M:37.2	0.947			
[22]	98	VAP	A/S	32	COL	66	47	56	0.34	82	52	0.03
							M:9.4	M:25.8	0.07			
							aOR: 6.5 (1.34–31.34) 0.02					
[23]	84	VAP + B	A/S+I/C	56	TG+I/C	28	NR	NR		NR	NR	
							M:14.3	M:64.3	0.007	NR	NR	
[24]	106	Pneumonia	C/S	35	CARB	46	71.4	29.3	0.003	NR	NR	
					TG	25		60	0.355			
							M:5.7	M:6.5				
								M:8				
[25]	130	VAP	C/S	66	TG	42	70	62	0.402	50	33	0.21
					C/S+TG	22	M:5	45	0.208	41	33	0.54
								M:12	0.295 *			
								M:27	0.231 **			
							aOR: 0.115 (0.015–0.89)		0.038			
[28]	47	VAP	A/S+MR	23	COL+MR	24	69.6	75	0.75	91.3	87.5	0.59
							M:39.13	M:41.67	0.99			
[29]	23	VAP	A/S+L	12	COL+L	11	83	27	0.007	75	100	NR
							M:41.66	M:81.81	0.04			
[26]	42	VAP	C/S+TG	21	TG	21	85.7	47.6	0.01	NR	NR	
[30]	28	VAP	A/S+nCOL	16	COL+nCOL	12	31.2	33.3	NS	43.7	12.5	0.37
							M:16.7	M:37.5	0.22			
[27]	180	VAP/HAP	A/SorC/S+COL	90	CARB+COL	90	NR	NR	0.658	NR	NR	
							M:51.1	M:55.6				
[31]	125	Pneumonia	S/D	63	COL	62	62	40	NR	86	61	NR
						M:19		M:32	0.0935 ¶			

VAP = Ventilator-associated pneumonia; B = bacteraemia; A/S = ampicillin/sulbactam; C/S = Cefoperazone/sulbactam; CARB = Carbapenem; I/C = imipenem/Cilastatin; COL = Colistin; nCOL = nebulised Colistin; MR= meropenem; TG = tigecycline; L = Levofloxacin; V = Various, i.e., 68.6% Carbapenem, 8.7% cephalosporins, 3.5% Fluoroquinolones; M = mortality; * = comparison between C/S and TG; ** = comparison between TG and C/S+TG; aOR = adjusted Odds Ratio; NR = not reported; ¶ = log rank p.

It is evident from the mentioned studies of the last decade that sulbactam not only retained significant clinical effectiveness against *A. baumannii* infections in an era in which antimicrobial resistance became worse on global scale but is also associated with lower mortality against comparator regimens. The new entry of durlobactam/sulbactam holds promise with less potential for nephrotoxicity. The availability of cefoperazone/sulbactam and the data from many Asian countries lend credibility to the role of sulbactam in the treatment of *A. baumannii* infections.

The present analysis is in full alignment with the IDSA guidelines [9] and the ESCMID guidelines [10] that suggest that when the isolates are susceptible to sulbactam one should consider it as a primary agent for the treatment of *A. baumannii* infections and indeed, interestingly, even if the strains are not susceptible to it. In the studies mentioned above, the sulbactam treatment arm was better or at least equal when compared with various agents, predominantly colistin and tigecycline, in the treatment of multidrug-resistant *A. baumannii* infections.

3.1.2. Cefiderocol

Cefiderocol is a siderophore cephalosporin which has FDA approval for the treatment of *A. baumannii* infections. The unique characteristic of this iron bearing cephalosporin has proved clinically important for Enterobacteriaceae and Pseudomonas aeruginosa as well. Data regarding *A. baumannii* are relatively few. We include the two randomised controlled trials (RCTs) as well as seven non-randomised, mainly retrospective observational, trials reporting on the effect of cefiderocol against a comparator antibiotic in Table 2. It is interesting that in the APEKS-NP study with 312 adult patients with VAP or health-care-associated Gram-negative pneumonia (including 47 due to *A. baumannii*) enrolled, cefiderocol was non-inferior to high-dose, extended-infusion meropenem in terms of all-cause mortality on day 14 [33].

Contradicting results were provided by the second RCT, the CREDIBLE-CR [34], regarding the subset of patients with *A. baumannii*, in which 19 out of 39 patients (49%) in the cefiderocol treatment arm died in comparison with 3 out of 17 (18%) in the best available treatment arm. This pointed to a double mortality rate in the cefiderocol treatment arm in patients with nosocomial pneumonia or bloodstream infection due to *A. baumannii*. Obviously, this is a very small sample size to evaluate the effect of any medication, especially as cefiderocol has been licenced by the FDA for *A. baumannii* infections. In addition, one must note that in Acinetobacter-infected patients, 26% had septic shock within a month before or even at randomisation in the cefiderocol group vs. 6% in the best available therapy group. Indeed, 81% of *A. baumanni*-infected patients that were randomised to cefiderocol were in the ICU at randomisation vs. 47% of those that received the comparator. The effect of baseline septic shock on mortality has been presented in a propensity-matched cohort study of 9000 patients with XDR infections [35], which showed a nine-fold higher mortality for patients with septic shock. One has to take this into consideration to evaluate the increased mortality in the cefiderocol arm shown in the CREDIBLE-CR study.

While the above two studies [33,34] are the only randomised studies until today, there have been a significant number of studies performed in recent years [36–42], and we present these data in Table 2. Most of these studies were conducted in Italy and are retrospective in nature. A total of 744 patients were included in the studies in the table [33,34,36–42] that addressed the therapeutic management of VAP and bloodstream carbapenem-resistant *A. baumannii* (CRAB) infections. When examining the mortality of patients that received a cefiderocol-containing regimen versus a comparator regimen that did not contain cefiderocol, mortality was observed in 152 out of the 367 patients (41.4%) that received cefiderocol versus 247 out of the 430 patients (57.4%) that received a comparator antibiotic. These findings suggest that cefiderocol could possibly be associated with lower all-cause mortality than previously thought. The data presented above regarding the potential role of cefiderocol in the treatment of *A. baumannii* infections cannot be ignored. There are conflicting suggestions by the IDSA [9] and ESCMID [10] guidelines regarding the use of cefiderocol in *A. baumannii* infections. Indeed, the IDSA suggests the use of cefiderocol as part of a combination regimen and when non tolerance to other therapeutic agents is present or when no other choices are available. On the other hand, there is a conditional recommendation against the use of cefiderocol in the ESCMID guidelines. We believe that, based on the literature, there is certainly a role for the use of cefiderocol. Obviously, cefiderocol-resistant *A. baumannii* strains are emerging [43]. Nevertheless, this agent retains a significant in vitro activity against *A. baumannii*, as shown by worldwide data [44–49]. Indeed, in a survey among 13 infectious disease specialists and 11 anaesthesiologists in Italy regarding the treatment of ICU patients in 87% of cases, cefiderocol was used as an empirical or targeted therapy when *A. baumannii* was the suspected pathogen [50].

The most used pharmacological combination when treating MDR *A. baumannii* infections using cefiderocol was fosfomycin (66.7%), followed by colistin (52.4%) and ampicillin/sulbactam (42.9%). The results of the GAME CHANGER trial are awaited to further evaluate the effectiveness of cefiderocol in the treatment of VAP and BSIs due to Gram-negative bacteria (including *A. baumannii*) [51].

3.1.3. Polymyxins

Polymyxins have been at the forefront of treatment for carbapenem-resistant *A. baumannii* infections for approximately the last 20 years [52]. The two polymyxins used are colistin (polymyxin E) and polymyxin B. The necessity of using them originated form the lack of suitable alternative antibiotics [53–61]. Polymyxins were used, though, before their re-emergence in the treatment of infections due to non-fermenters in patients with cystic fibrosis. While nephrotoxicity is a relative concern of their use, this is usually reversible and should not deter physicians from prescribing these drugs. Although most of the studies reporting on the clinical use of colistin in the past were non-randomised, their effectiveness and safety are undisputable. We summarise non-randomised studies performed in the last decade regarding the use of colistin [62–71] in Table 3 and randomised comparative studies [72,73] in Table 4. It is obvious from the data presented that colistin retains significant clinical effectiveness and favourable outcomes on decreased mortality even as a monotherapy. Combinations with sulbactam or tigecycline seem to confer a benefit. Indeed, in the non-randomised studies mortality was similar in patients receiving colistin as a monotherapy (47.1%, 302 out of 642 patients) in comparison to combination therapy (44.9%, 376 out of 836 patients). Mortality was also similar in the randomised studies when comparing patients receiving colistin monotherapy (44.6%, 167 out of 374 patients) with patients receiving combination therapy (46.3%, 172 patients out of 371).

Non parenteral administration of colistin has also been reported. Nebulised colistin has been used mainly as an adjunctive to intravenous treatment and rarely as a monotherapy without providing any significant benefit in the majority of studies, although data are scarce and mostly older than the timeframe we elected to evaluate [74,75]. While a potential drawback is bronchospasm in some patients, the medication has also been used in children in its aerosolised form, without any major safety concern in patients with and without cystic fibrosis. Intrathecal or intraventricular colistin has a significant role in the treatment of neurosurgical infections [76].

Among polymyxins, polymyxin E (colistin) is used more frequently. Nevertheless, there are data in the literature regarding the use of polymyxin B. Polymyxin B is mostly used in North and South America and Asia. While, generally, polymyxin B has been shown in various studies to be less nephrotoxic than colistin, a recent meta-analysis showed a similar nephrotoxicity. We identified a significant number of studies during the last decade pointing to the clinical and microbiological effectiveness and safety of polymyxin B as a treatment for *A. baumannii* infections [77–99]. Indeed, the agent is equally effective with colistin and relatively safer in regard to nephrotoxicity [100–102].

Table 2. Summary of comparative studies examining efficacy of cefiderocol in patients with *A. baumannii* infections.

Study (Year)	No. of Evaluable	Infection ¶	Antibacterial Regimens No. Pts				Cure and/or Improvement (%) Mortality (%)			Bacteriological Eradication (%)		
			Comparator 1		Comparator 2		Comparator 1	Comparator 2	*p*-Value	Comparator 1	Comparator 2	*p*-Value
[33]	47	Pneumonia	CFD	23	MR	24	52	58	NS	39	33	NS
							M:19	M:22	NS			
[34]	54	Pneumonia/B/LTI	CFD	37	BAT	17	43	53	NS	27	29	NS
							M:49	M:18	NR			
[36]	107	B:27 LRTI:14 ALL COVID-19	CFD	42	COL−R	65	40	36	0.45	28	21	0.24
							M:55	M:58	0.70			
[37]	124	B:79 VAP:35	CFD−R	47	COL−R	28	NR	NR				
							M:34%	M:55.8	0.018 *	82.6	93.2	0.079
[38]	111	B:53 P:47	CFD−R	60	COL−R	51	73	67	0.44	43	41	0.82
							M:51	M:37				
[39]	118	B	CFD−R	43	COL−R	75	NR	NR		NR	NR	
							M:40	M:59	0.045			
[40]	121	VAP ALL COVID-19	CFD−R	55	NON CFD−R	66	NR	NR		47	69	0.038
							M:44	M:67	0.011			
[41]	90	VAP	CFD+Ncol	40	COL+nCOL	50	75	52	0.02	70	40	0.003
							M:35	M:52	NS			
[42]	73	Bacteremic VAP ALL COVID-19	CFD−R	19	COL−R	54	NR	NR		NR	NR	
							M:31.5	M:98.1	<0.001			

Abbreviations: No. = number; Pts = patients; M = mortality; ¶ = numbers in the infection column are absolute numbers of patients with the specific type of infection mentioned; VAP = Ventilator-associated pneumonia; *p* = pneumonia; LRTI = lower respiratory tract infection; B = bloodstream infection; CFD = cefiderocol; CFD-R = cefiderocol-containing regimen; COL-R = colistin-containing regimens, i.e., combination regimens ± monotherapy; COL = Colistin; nCOL = nebulised colistin; * = regarding mortality of bacteraemia only; MR= meropenem; BAT = best available treatment; NR = not reported; NS = non-significant.

Table 3. Summary of non-randomised comparative studies examining efficacy of polymyxins in patients with *A. baumannii* infections.

Study (Year)	No. of Evaluable	Infection	Antibacterial Regimens No. Pts				Cure and/or Improvement (%)			Bacteriological Eradication (%)		
			Comparator 1		Comparator 2		Comparator 1	Comparator 2	*p*-Value	Comparator 1	Comparator 2	*p*-Value
[62]	250	B	COL	36	COL+V	214	30.6	46.3	0.19	55.6	79.9	0.001
							M:44.5	M:31.8	0.14			
[63]	89	VAP	COL	57	COL+S	32	29.8	40	0.50	72.3	85.7	0.28
							M:51.9	M:73	0.53			
[64]	79	VAP SSTI, UTI	COL	46	COL+V	23	NR	NR		NR	NR	
							M:26.1	M:26.08	NS			
[65]	55	B	COL+C	26	COL+TG	29	NR	NR		100	82	
							M:15	M:35	0.1	NR	NR	
[66]	70	VAP	COL	13	COL+S	20	76.5	55	0.53	52.9	63.6	0.16
					COL+C	33		63.6	0.35		60	0.23
							M:41.2	M:70	0.21			
								M:48.5	0.5			
[67]	107	B	COL	36	NON−COL	71	77.1	77.1	0.45	69	83	0.13
							M:47.2	M:52.77	0.36			
[68]	118	B	COL	76	COL+TG	42	NR	NR		NR	NR	
							M:22	M:24	1			
[69]	39	VAP	COL	19	COL+A/S	20	15.8	70	0.001	33.3	71.4	0.19
							M:63	M:50	NS			
[70]	71	B	COL	40	COL+MR	31	40	61.3	0.07	NR	NR	
							M:47.5	M:25.8	0.08			
[71]	617	NP	COL	293	NON−COL	324	aOR: 1.27 (0.92–1.75) ¶			aOR: 3.44 (2.36–5.02) ¶		
							M:54.3 *	M:53.7 *	0.87			

B = bloodstream infection; VAP = Ventilator-associated pneumonia; SSTI = skin and soft tissue infection; NP = nosocomial pneumonia; COL = Colistin; V = Various; S= sulbactam; TG = tigecycline; C = Carbapenem; NON-COL= regimen not containing colistin; A/S = ampicillin/sulbactam; MR= meropenem; M = mortality; aOR = adjusted Odds Ratio; NR = not reported; ¶ = in favour of colistin. * Patients with multiple organ failure and SOFA score more than 7 treated with colistin-based regimen had better survival (aOR: 0.38) than non-colistin regimens (aOR: 0.52), $p = 0.011$.

Table 4. Summary of randomised comparative studies examining efficacy of polymyxins in patients with *A. baumannii* infections.

Study (Year)	No. of Evaluable	Infection	Antibacterial Regimens No Pts				Cure and/or Improvement (%)			Bacteriological Eradication (%)		
			Comparator 1		Comparator 2		Comparator 1	Comparator 2	*p*-Value	Comparator 1	Comparator 2	*p*-Value
[73]	312	VAP B, UTI	COL	151	COL+M	161	21	27	0.643	69	65	0.489
					M:46	M:52	0.404					
[72]	413	VAP, B W, UTI	COL	213	COL+M	210	32	40	NS	63	57	NS
					M:46	M:42	NS					

VAF = Ventilator-associated pneumonia, B = bloodstream infection, W = wound infection, UTI = urinary tract infection, COL = Colistin, M = meropenem, M = mortality, NS = non-significant.

3.1.4. Tetracyclines and Glycylcyclines

Tigecycline retains significant in vitro activity against *A. baumannii* [103] and has been used in the treatment of infections caused by this organism. However, tigecycline, as with all antibiotics, is not immune to antimicrobial resistance mechanisms [104]. In addition, pharmacokinetics regarding blood levels have shown that the levels achieved are not adequate for the treatment of *A. baumannii* bacteraemia [105]. Furthermore, the same holds true for the levels in the lungs [106,107]. It is therefore not surprising that higher doses (indeed, double doses) are suggested for use in clinical practice [108,109]. Therefore, high-dose tigecycline has been suggested (200 mg loading dose, followed by 100 mg bid) and indeed has been found more effective than standard-dose tigecycline [108,109]. We identified seven retrospective studies [110–116] during the last decade (Table 5) that compared tigecycline vs. a comparator antibiotic, colistin in five and sulbactam in two. In all, tigecycline was administered in the standard dosing, i.e., a 100 mg iv initially and then 50 mg twice daily. In total, 182 out of the 374 patients that received tigecycline died (48.6%) versus 200 out of the 508 (39.3%) in the comparator arm. Breaking it down, mortality was better in the tigecycline arm in one study, worse in two and without statistical difference in three. One must note that in one of these studies, which only enrolled patients with bloodstream infections [116], mortality was 22% higher in the tigecycline arm.

3.1.5. Minocycline

There are relatively few data points regarding the use of minocycline against *A. baumannii* infections. Minocycline is an antibiotic belonging to the tetracyclines group and has been used in the therapeutic arena for over 6 decades. Its use is indicated in acne and in a variety of infections cause by aerobic Gram-positive and Gram-negative microorganisms, intracellular microorganisms and, interestingly, also against *A. baumannii* [117,118]. In a meta-analysis, minocycline was reported to have been used in the treatment of *A. baumannii* infection in 228 patients [119]. The clinical and microbiological success rates reported following minocycline treatment were 72.6% and 60.2%, respectively. Nevertheless, in the vast majority (~92% of these patients) it was part of a combination regimen, mainly with carbapenems or colistin. Only a handful of studies have been conducted in the last decade [120–124] that support the use of minocycline in the treatment of A. baumanni infections. Susceptibility to minocycline seems to be correlated with the absence of the TeTB gene [125].

Table 5. Summary of comparative studies examining efficacy of tigecycline in patients with *A. baumannii* infections.

Study (Year)	No. of Evaluable	Infection	Antibacterial Regimens No. Pts				Cure and/or Improvement (%)			Bacteriological Eradication (%)		
			Comparator 1		Comparator 2		Comparator 1	Comparator 2	p-Value	Comparator 1	Comparator 2	p-Value
[110]	88	VAP	TG	44	COL	44	NR	NR		NR	NR	
							M:60.7	M:44	0.04			
[111]	55	VAP, B W, UTI	TG	16	COL	39	43.8	48.7	0.737	12.5	46.2	0.049
							M:56.3	M:43.6	0.393			
[112]	79	VAP	TG+COL	19	COL+I	60	NR	NR		NR	NR	19
						M:26.3	M:53.3	0.04				
[113]	70	VAP±B	TG−R	30	COL−R	40	47	48	0.95	23	30	0.54
						M:33	M:30	0.77				
[114]	168	VAP	TG−R	84	SUL−R	84	66.7	66.7	1	33.3	63.5	<0.0001
						M:25	M:17.9	0.25				
[115]	212	VAP	nCOL+TG	106	nCOL	106	NR *	NR *		NR	NR	
						M:34	M:22.6	0.17				
[116]	210	B	TG	75	C/S	135	NR	NR		NR	NR	
							M:51.9	M:29.3	0.001			

VAP = Ventilator-associated pneumonia, B = bloodstream infection, W = wound infection, UTI = urinary tract infection, I = imipenem/Cilastatin, TG = tigecycline, COL = Colistin, nCOL = nebulised Colistin, M = mortality, NR = not reported, NR * = lack of TGC therapy (aHR = 0.52; 95% CI = 0.27–1.00; p = 0.049) adversely influenced the 14-day clinical cure.

3.1.6. Trimethoprim/Sulfamethoxazole (TMP/SMX)

There is only one retrospective cohort study, by Raz et al., that included 53 patients receiving TMP/SMX for severe infections caused by CRAB who were matched with 83 patients treated with colistin or ampicillin/sulbactam [126]. A variety of infections were present in these patients: pneumonia (71.3%) was the most frequent, followed by skin and soft tissue infections, UTIs and central line-associated bacteraemia. All-cause 30-day mortality was lower with TMP/SMX compared with the comparator antibiotics among all patients (24.5% vs. 38.6%, $p = 0.09$) and in the propensity score-matched subgroup (29% vs. 55.2%, $p = 0.04$).

3.1.7. Fosfomycin and Rifampicin

Fosfomycin and rifampicin are antibiotics that have no satisfactory clinical effectiveness against *A. baumannii*. However, there are scarce data in the literature supporting their role as part of combination regimens with other antimicrobial agents.

Sirijatuphat and Thamlikitkul randomised 94 patients infected with carbapenem-resistant *A. baumannii* to receive colistin alone or colistin plus fosfomycin for 7 to 14 days [127]. A survival analysis showed no significant difference between the patients who received combination therapy and monotherapy. However, microbiological eradication rates in the combination group were significantly higher than those in the monotherapy group, 100% versus 81.2%, at the end of study treatment.

Russo et al. performed a prospective study including 44 patients with hospital-acquired pneumonia due to MDR-*A. baumannii* strains [128] that were treated with fosfomycin combinations. The following combinations were administered: fosfomycin plus colistin in 11 (25%) patients, fosfomycin plus carbapenem plus tigecycline in 8 (18.2%), fosfomycin plus colistin plus tigecycline in 7 (15.9%), fosfomycin plus rifampin in 7 (15.9%), fosfomycin plus tigecycline in 6 (13.6%), fosfomycin plus carbapenem in 3 (6.8%), and fosfomycin plus aminoglycoside in 2 (4.5%). Cox regression analysis showed that a fosfomycin-containing regimen was associated with 30-day survival (HR 0.04, CI 95% 0.01–0.13, $p < 0.001$).

Regarding rifampicin, we identified only one RCT in the time period we selected to evaluate [129] with a very small sample size of nine patients with pneumonia caused by colistin-resistant *A. baumannii*. In this study, patients were randomised to colistin/rifampicin combination therapy or colistin monotherapy. Unsurprisingly, this study failed to show any difference in clinical and microbiological outcomes. Prior reports, such as those by Durante-Magnoni et al. [61] and Aydemir et al. [130], also questioned the clinical significance of adding rifampicin to colistin.

3.2. Newer Beta-Lactamase Inhibitors

There are no clinical trials regarding the clinical effectiveness of the newer beta-lactamase inhibitors (avibactam, relebactam, varbobactam).

4. Discussion

From the studies evaluated, it is suggested that for MDR/XDR *A. baumannii* isolates, which usually are resistant to beta-lactams (including carbapenems) and quinolones, one has to resort to high-dose sulbactam (6–9 g daily), colistin (intravenous ± nebulised), or polymyxin B or high-dose tigecycline (i.e., 100 mg bid).

High-dose sulbactam was initially suggested in 2008. The data from the last decade presented herein (Table 1) corroborate its effectiveness in clinical practice. It is evident from the mentioned studies from the last decade that sulbactam not only retained significant clinical effectiveness against *A. baumannii* infections in an era in which antimicrobial resistance became worse on a global scale but is also associated with lower mortality against comparator regimens. Indeed, mortality was approximately 20 percent less in patients that received sulbactam than the comparator. The new entry of durlobactam/sulbactam holds promise with less potential for nephrotoxicity. The availability of cefoperazone/sulbactam and the data from many Asian countries lend credibility to the role of sulbactam in the treatment of *A. baumannii* infections.

The present analysis is in full alignment with the IDSA guidelines [9] and the ESCMID guidelines [10] that suggest that when the isolates are susceptible to sulbactam one should consider it as a primary agent for the treatment of *A. baumannii* infections and indeed, interestingly, even if the strains are not susceptible to it. In the studies mentioned above, the sulbactam treatment arm was better or at least equal when compared with various agents, predominantly colistin and tigecycline, in the treatment of multidrug-resistant *A. baumannii* infections.

Therefore, the ESCMID and IDSA guidelines that view sulbactam as a first-line option for MDR/XDR *A. baumannii* isolates are, in our opinion, more than justified. As sulbactam is usually found in parenteral form in various combinations with other antimicrobials, such as ampicillin (ampicillin 1 g combined with 0.5 g of sulbactam, ampicillin 2 g combined with 1 g of sulbactam), cefoperazone (cefoperazone 1 g combined with 0.5 g of sulbactam, cefoperazone 1 g combined with 1 g of sulbactam or cefoperazone 2 g combined with 1 g of sulbactam) or durlobactam, one has to calculate the amount of sulbactam necessary to administer to the patient.

Based on the data from the literature presented, polymyxin E (colistin) and polymyxin B retain their crucial role in the treatment of severe *A. baumannii* infections. Indeed, in the majority of studies, polymyxins fared equally well as monotherapies when compared with combination regimens (Tables 3 and 4). While the vast majority of studies in the literature do not support a combination regimen of these agents, both the IDSA and the European guidelines advise in favour of such a combination. Indeed, the IDSA and European guidelines state that it is preferable to use combinations of antimicrobial agents to which susceptibility testing has been performed and found relevant. We concur, in view of the severity of these infections (patients are usually critically ill), the possibility of the emergence of resistance, the high bacterial quantities that predate targeted antibiotic therapy (i.e., prior to bacterial culture and susceptibility results) and because of suboptimal doses of colistin (at times, the prescribed dose was half of what should be given), that the use of combination regimens is justified and safe. However, further data from high-quality RCTs are much needed in order to corroborate or negate the theoretical advantage of a combination regimen. While polymyxin B is favoured in general in the guidelines, based on our experience [57] and the data from the literature [62–73] we do not believe that colistin is inferior to polymyxin B in terms of clinical effectiveness. We acknowledge the smaller degree of nephrotoxicity associated with polymyxin B. Nephrotoxicity is a potential adverse event of colistin treatment but it is usually reversible and should not prevent clinicians from prescribing the drug [131]. Adjustments of colistin dosing have been agreed upon in consensus meetings of scientific associations and provide useful guidance for adjustments based on creatinine clearance measurements [132]. We therefore believe that colistin should not be stricken so easily from the therapeutic armamentarium and just preferred for use in UTIs over polymyxin B, especially as minocycline and polymyxin B are not available in parenteral forms worldwide.

Regarding tigecycline, one must be aware that the standard dose of tigecycline (i.e., a 100 mg loading dose iv followed by a 50 mg bid) does not achieve adequate levels in the blood and possibly in the respiratory system to combat MDR/XDR *A. baumannii* [104–107], and this translates to worse clinical outcomes in clinical practice in comparison to comparators. Therefore, based on the findings of the comparative studies presented in Table 4, we suggest that standard dosing of tigecycline should be avoided for the treatment of *A. baumannii* infections and only high-dose tigecycline should be employed. On the other hand, high-dose tigecycline (a 200 mg loading dose followed by a 100 mg bid) provides a useful antimicrobial agent in combination with one of the other agents active against carbapenem-resistant *A. baumannii* (i.e., colistin or polymyxin B or high-dose sulbactam) [108,109].

A crucial issue is also the use of cefiderocol. The European guidelines advise against the use of cefiderocol as a treatment for *A. baumannii* infections, while the IDSA guidelines advise its use on a conditional basis provided that there are no alternatives or there is intolerance to other antimicrobials. The IDSA guidelines advise that cefiderocol, if used, is to be part of a combination regimen with caution until and if more favourable data are available. Their recommendation is mainly based on the data provided by the RCT from Basseti et al., which had a very small number (54) of patients with *A. baumannii* infections that disfavoured cefiderocol treatment, and those by the APEKS-NP, which showed no difference. We could not agree more with the IDSA guidelines on this issue that caution is needed. Nevertheless, there is already a significant number of trials (although non-randomised) presented herein that are in favour of cefiderocol. Indeed, the comparative studies by Pascale et al. [36], Falcone et al. [37]., Mazzitelli et al. [38], Bavaro et al. [39], Rando et al. [40], Dalfino et al. [41] and Russo et al. [42] included 637 patients, and 306 of them received cefiderocol for *A. baumannii* infections with favourable clinical outcomes.

Aminoglycosides are proposed as first-line agents by both the IDSA and ESCMID in the treatment of carbapenem-resistant *A. baumannii*. Nevertheless, there are no data in the current literature to support aminoglycoside monotherapy in severe infections such as VAP and bacteraemia. Therefore, they must not be given as a monotherapy for these severe *A. baumannii* infections. Aminoglycosides can be used in these types of infections as a part of combination regimens.

Regarding fosfomycin and rifampicin, no convincing clinical evidence exists regarding their use, and it seems that their role in the treatment of *A. baumannii* infections remains to be clarified by larger trials.

5. Conclusions

In conclusion, one hopes for additional trials, especially randomised trials, that will further clarify the ever-lasting challenge of *A. baumannii* infections' treatment. While prevention through infection control measures is of paramount significance, a clear understanding of the current literature is mandatory to provide suitable therapeutic management to critically ill patients with *A. baumannii* infections. Hence, based on our review, one has to resort to at least one of the following, sulbactam, polymyxins E or B, tigecycline or aminoglycosides, against multidrug-resistant (MDR) or extensively drug-resistant (XDR) *A. baumannii* infections. Furthermore, cefiderocol is a useful addition in the fight against *A. baumannii* infections.

Author Contributions: Conceptualisation, P.R., P.P. and G.S.; methodology, P.R. and P.P.; validation, P.R., P.P., C.K. and G.S.; investigation, P.R., P.P. and C.K.; resources, C.K. and G.S.; data curation, P.R. and P.P.; writing—original draft preparation, P.R. and P.P.; writing—review and editing, P.R., P.P., C.K. and G.S.; visualisation, P.R., P.P. and G.S.; supervision, P.R., P.P. and G.S.; project administration, P.R. and G.S. All authors have read and agreed to the published version of the manuscript.

Funding: This research received no external funding.

Data Availability Statement: Data are available through the Pubmed and Scopus databases.

Conflicts of Interest: The authors declare no conflicts of interest.

References

1. Pogue, J.M.; Zhou, Y.; Kanakamedala, H.; Cai, B. Burden of illness in carbapenem-resistant *Acinetobacter baumannii* infections in US hospitals between 2014 and 2019. *BMC Infect Dis.* **2022**, *22*, 36. [CrossRef]
2. Peleg, A.Y.; Seifert, H.; Paterson, D.L. *Acinetobacter baumannii*: Emergence of a successful pathogen. *Clin. Microbiol. Rev.* **2008**, *21*, 538–582. [CrossRef]
3. Iovleva, A.; Mustapha, M.M.; Griffith, M.P.; Komarow, L.; Luterbach, C.; Evans, D.R.; Cober, E.; Richter, S.S.; Rydell, K.; Arias, C.A.; et al. Carbapenem-Resistant *Acinetobacter baumannii* in U.S. Hospitals: Diversification of Circulating Lineages and Antimicrobial Resistance. *mBio* **2022**, *13*, e0275921. [CrossRef]
4. Xu, S.; Li, Y.; Xu, X.; Su, J.; Zhu, D.; Hu, F.; Wang, M. A Case-Control Study: Clinical Characteristics of Nosocomial Bloodstream Infections Versus Non-Bloodstream Infections of *Acinetobacter* spp. *Clin. Infect. Dis.* **2018**, *67*, S189–S195. [CrossRef]
5. Zilberberg, M.D.; Kollef, M.H.; Shorr, A.F. Secular trends in *Acinetobacter baumannii* resistance in respiratory and blood stream specimens in the United States, 2003 to 2012: A survey study. *J. Hosp. Med.* **2016**, *11*, 21–26. [CrossRef] [PubMed]

6. Magiorakos, A.P.; Srinivasan, A.; Carey, R.B.; Carmeli, Y.; Falagas, M.E.; Giske, C.G.; Harbarth, S.; Hindler, J.F.; Kahlmeter, G.; Olsson-Liljequist, B.; et al. Multidrug-resistant, extensively drug-resistant and pandrug-resistant bacteria: An international expert proposal for interim standard definitions for acquired resistance. *Clin. Microbiol. Infect.* **2012**, *18*, 268–281. [CrossRef] [PubMed]
7. Kadri, S.S.; Adjemian, J.; Lai, Y.L.; Spaulding, A.B.; Ricotta, E.; Prevots, D.R.; Palmore, T.N.; Rhee, C.; Klompas, M.; Dekker, J.P.; et al. Difficult-to-Treat Resistance in Gram-negative Bacteremia at 173 US Hospitals: Retrospective Cohort Analysis of Prevalence, Predictors, and Outcome of Resistance to All First-line Agents. *Clin. Infect. Dis.* **2018**, *67*, 1803–1814. [CrossRef] [PubMed]
8. Mohd Sazlly Lim, S.; Zainal Abidin, A.; Liew, S.M.; Roberts, J.A.; Sime, F.B. The global prevalence of multidrug-resistance among *Acinetobacter baumannii* causing hospital-acquired and ventilator-associated pneumonia and its associated mortality: A systematic review and meta-analysis. *J. Infect.* **2019**, *79*, 593–600. [CrossRef] [PubMed]
9. Tamma, P.D.; Aitken, S.L.; Bonomo, R.A.; Mathers, A.J.; van Duin, D.; Clancy, C.J. Infectious Diseases Society of America Guidance on the Treatment of AmpC β-Lactamase-Producing Enterobacterales, Carbapenem-Resistant *Acinetobacter baumannii*, and *Stenotrophomonas maltophilia* Infections. *Clin. Infect. Dis.* **2022**, *74*, 2089–2114. [CrossRef] [PubMed]
10. Paul, M.; Carrara, E.; Retamar, P.; Tängdén, T.; Bitterman, R.; Bonomo, R.A.; de Waele, J.; Daikos, G.L.; Akova, M.; Harbarth, S.; et al. European Society of Clinical Microbiology and Infectious Diseases (ESCMID) guidelines for the treatment of infections caused by multidrug-resistant Gram-negative bacilli (endorsed by European society of intensive care medicine). *Clin. Microbiol. Infect.* **2022**, *28*, 521–547. [CrossRef] [PubMed]
11. Jones, R.N.; Wilson, H.W.; Thornsberry, C.; Barry, A.L. In vitro antimicrobial activity of cefoperazone-sulbactam combinations against 554 clinical isolates including a review and beta-lactamase studies. *Diagn. Microbiol. Infect. Dis.* **1985**, *3*, 489–499. [CrossRef] [PubMed]
12. Eliopoulos, G.M.; Klimm, K.; Ferraro, M.J.; Moellering, R.C. In vitro activity of cefoperazone-sulbactam combinations against cefoperazone-resistant clinical bacterial isolates. *Eur. J. Clin. Microbiol. Infect. Dis.* **1989**, *8*, 624–626. [CrossRef] [PubMed]
13. Corbella, X.; Ariza, J.; Ardanuy, C.; Vuelta, M.; Tubau, F.; Sora, M.; Pujol, M.; Gudiol, F. Efficacy of sulbactam alone and in combination with ampicillin in nosocomial infections caused by multiresistant *Acinetobacter baumannii*. *J. Antimicrob. Chemother.* **1998**, *42*, 793–802. [CrossRef] [PubMed]
14. Cisneros, J.M.; Reyes, M.J.; Pachón, J.; Becerril, B.; Caballero, F.J.; García-Garmendía, J.L.; Ortiz, C.; Cobacho, A.R. Bacteremia due to *Acinetobacter baumannii*: Epidemiology, clinical findings, and prognostic features. *Clin. Infect. Dis.* **1996**, *22*, 1026–1032. [CrossRef] [PubMed]
15. Jellison, T.K.; Mckinnon, P.S.; Rybak, M.J. Epidemiology, resistance, and outcomes of *Acinetobacter baumannii* bacteremia treated with imipenem-cilastatin or ampicillin-sulbactam. *Pharmacotherapy* **2001**, *21*, 142–148. [CrossRef] [PubMed]
16. Wood, G.C.; Hanes, S.D.; Croce, M.A.; Fabian, T.C.; Boucher, B.A. Comparison of ampicillin-sulbactam and imipenem-cilastatin for the treatment of *Acinetobacter* ventilator-associated pneumonia. *Clin. Infect. Dis.* **2002**, *34*, 1425–1430. [CrossRef]
17. Levin, A.S.; Levy, C.E.; Manrique, A.E.; Medeiros, E.A.; Costa, S.F. Severe nosocomial infections with imipenem-resistant *Acinetobacter baumannii* treated with ampicillin/sulbactam. *Int. J. Antimicrob. Agents.* **2003**, *21*, 58–62. [CrossRef]
18. Smolyakov, R.; Borer, A.; Riesenberg, K.; Schlaeffer, F.; Alkan, M.; Porath, A.; Rimar, D.; Almog, Y.; Gilad, J. Nosocomial multi-drug resistant *Acinetobacter baumannii* bloodstream infection: Risk factors and outcome with ampicillin-sulbactam treatment. *J. Hosp. Infect.* **2003**, *54*, 32–38. [CrossRef]
19. Oliveira, M.S.; Prado, G.V.; Costa, S.F.; Grinbaum, R.S.; Levin, A.S. Ampicillin/sulbactam compared with polymyxins for the treatment of infections caused by carbapenem-resistant *Acinetobacter* spp. *J. Antimicrob. Chemother.* **2008**, *61*, 1369–1375. [CrossRef]
20. Betrosian, A.P.; Frantzeskaki, F.; Xanthaki, A.; Douzinas, E.E. Efficacy and safety of high-dose ampicillin/sulbactam vs. colistin as monotherapy for the treatment of multidrug resistant *Acinetobacter baumannii* ventilator-associated pneumonia. *J. Infect.* **2008**, *56*, 432–436. [CrossRef]
21. Lin, H.S.; Lee, M.H.; Cheng, C.W.; Hsu, P.C.; Leu, H.S.; Huang, C.T.; Ye, J.J. Sulbactam treatment for pneumonia involving multidrug-resistant *Acinetobacter calcoaceticus-Acinetobacter baumannii* complex. *Infect. Dis.* **2015**, *47*, 370–378. [CrossRef]
22. Zalts, R.; Neuberger, A.; Hussein, K.; Raz-Pasteur, A.; Geffen, Y.; Mashiach, T.; Finkelstein, R. Treatment of Carbapenem-Resistant *Acinetobacter baumannii* Ventilator-Associated Pneumonia: Retrospective Comparison between Intravenous Colistin and Intravenous Ampicillin-Sulbactam. *Am. J. Ther.* **2016**, *23*, e78–e85. [CrossRef]
23. Jean, S.S.; Hsieh, T.C.; Hsu, C.W.; Lee, W.S.; Bai, K.J.; Lam, C. Comparison of the clinical efficacy between tigecycline plus extended-infusion imipenem and sulbactam plus imipenem against ventilator-associated pneumonia with pneumonic extensively drug-resistant *Acinetobacter baumannii* bacteremia, and correlation of clinical efficacy with in vitro synergy tests. *J. Microbiol. Immunol. Infect.* **2016**, *49*, 924–933.
24. Li, P.; Wang, X.; Wang, W.; Zhao, X. Comparison of the efficacies of three empirically-selected antibiotics for treating *Acinetobacter baumannii* pulmonary infection: Experience from a teaching hospital in China. *Int. J. Clin. Pharmacol. Ther.* **2017**, *55*, 588–593. [CrossRef]
25. Li, Y.; Xie, J.; Chen, L.; Meng, T.; Liu, L.; Hao, R.; Dong, H.; Wang, X.; Dong, Y. Treatment efficacy of tigecycline in comparison to cefoperazone/sulbactam alone or in combination therapy for carbapenem-resistant *Acinetobacter baumannii* infections. *Pak. J. Pharm. Sci.* **2020**, *33*, 161–168. [PubMed]
26. Qin, Y.; Zhang, J.; Wu, L.; Zhang, D.; Fu, L.; Xue, X. Comparison of the treatment efficacy between tigecycline plus high-dose cefoperazone-sulbactam and tigecycline monotherapy against ventilator-associated pneumonia caused by extensively drug-resistant *Acinetobacter baumannii*. *Int. J. Clin. Pharmacol. Ther.* **2018**, *56*, 120–129. [CrossRef] [PubMed]

27. Ungthammakhun, C.; Vasikasin, V.; Changpradub, D. Clinical Outcomes of Colistin in Combination with Either 6-G Sulbactam or Carbapenems for the Treatment of Extensively Drug-Resistant *Acinetobacter baumannii* Pneumonia with High MIC to Sulbactam, A Prospective Cohort Study. *Infect. Drug Resist.* **2019**, *12*, 2899–2904. [CrossRef] [PubMed]
28. Khalili, H.; Shojaei, L.; Mohammadi, M.; Beigmohammadi, M.T.; Abdollahi, A.; Doomanlou, M. Meropenem/colistin versus meropenem/ampicillin-sulbactam in the treatment of carbapenem-resistant pneumonia. *J. Comp. Eff. Res.* **2018**, *7*, 901–911. [CrossRef] [PubMed]
29. Mosaed, R.; Haghighi, M.; Kouchak, M.; Miri, M.M.; Salarian, S.; Shojaei, S.; Javadi, A.; Taheri, S.; Nazirzadeh, P.; Foroumand, M.; et al. Interim Study: Comparison of Safety and Efficacy of Levofloxacin Plus Colistin Regimen with Levofloxacin Plus High Dose Ampicillin/Sulbactam Infusion In Treatment of Ventilator-Associated Pneumonia Due To Multi Drug Resistant *Acinetobacter*. *Iran. J. Pharm. Res.* **2018**, *17*, 206–213. [PubMed]
30. Pourheidar, E.; Haghighi, M.; Kouchek, M.; Miri, M.M.; Shojaei, S.; Salarian, S.; Hassanpour, R.; Sistanizad, M. Comparison of Intravenous Ampicillin-sulbactam Plus Nebulized Colistin with Intravenous Colistin Plus Nebulized Colistin in Treatment of Ventilator Associated Pneumonia Caused by Multi Drug Resistant *Acinetobacter baumannii*: Randomized Open Label Trial. *Iran. J. Pharm. Res.* **2019**, *18*, 269–281. [PubMed]
31. Kaye, K.S.; Shorr, A.F.; Wunderink, R.G.; Du, B.; Poirier, G.E.; Rana, K.; Miller, A.; Lewis, D.; O'Donnell, J.; Chen, L.; et al. Efficacy and safety of sulbactam-durlobactam versus colistin for the treatment of patients with serious infections caused by *Acinetobacter baumannii-calcoaceticus* complex: A multicentre, randomised, active-controlled, phase 3, non-inferiority clinical trial (ATTACK). *Lancet Infect. Dis.* **2023**, *23*, 1072–1084. [PubMed]
32. Ungthammakhun, C.; Vasikasin, V.; Changpradub, D. A Randomized Controlled Trial of Colistin Combined with Sulbactam: 9 g per Day versus 12 g per Day in the Treatment of Extensively Drug-Resistant *Acinetobacter baumannii* Pneumonia: An Interim Analysis. *Antibiotics* **2022**, *11*, 1112. [CrossRef] [PubMed]
33. Wunderink, R.G.; Matsunaga, Y.; Ariyasu, M.; Clevenbergh, P.; Echols, R.; Kaye, K.S.; Kollef, M.; Menon, A.; Pogue, J.M.; Shorr, A.F.; et al. Cefiderocol versus high-dose, extended-infusion meropenem for the treatment of Gram-negative nosocomial pneumonia (APEKS-NP): A randomised, double-blind, phase 3, non-inferiority trial. *Lancet Infect. Dis.* **2021**, *21*, 213–225. [CrossRef] [PubMed]
34. Bassetti, M.; Echols, R.; Matsunaga, Y.; Ariyasu, M.; Doi, Y.; Ferrer, R.; Lodise, T.P.; Naas, T.; Niki, Y.; Paterson, D.L.; et al. Efficacy and safety of cefiderocol or best available therapy for the treatment of serious infections caused by carbapenem-resistant Gram-negative bacteria (CREDIBLE-CR): A randomised, open-label, multicentre, pathogen-focused, descriptive, phase 3 trial. *Lancet Infect. Dis.* **2021**, *21*, 226–240. [CrossRef] [PubMed]
35. Kadri, S.S.; Strich, J.R.; Swihart, B.J.; Hohmann, S.; Dekker, J.P.; Palmore, T.; Bonne, S.; Freeman, B.; Raybould, J.; Shah, N.G.; et al. Attributable mortality from extensively drug-resistant gram-negative infections using propensity-matched tracer antibiotic algorithms. *Am. J. Infect. Control* **2019**, *47*, 1040–1047. [CrossRef] [PubMed]
36. Pascale, R.; Pasquini, Z.; Bartoletti, M.; Caiazzo, L.; Fornaro, G.; Bussini, L.; Volpato, F.; Marchionni, E.; Rinaldi, M.; Trapani, F.; et al. Cefiderocol treatment for carbapenem-resistant *Acinetobacter baumannii* infection in the ICU during the COVID-19 pandemic: A multicentre cohort study. *JAC-Antimicrob. Resist.* **2021**, *3*, dlab174. [CrossRef]
37. Falcone, M.; Tiseo, G.; Leonildi, A.; Della Sala, L.; Vecchione, A.; Barnini, S.; Farcomeni, A.; Menichetti, F. Cefiderocol-Compared to Colistin-Based Regimens for the Treatment of Severe Infections Caused by Carbapenem-Resistant *Acinetobacter baumannii*. *Antimicrob. Agents Chemother.* **2022**, *66*, e0214221. [CrossRef]
38. Mazzitelli, M.; Gregori, D.; Sasset, L.; Trevenzoli, M.; Scaglione, V.; Lo Menzo, S.; Marinello, S.; Mengato, D.; Venturini, F.; Tiberio, I.; et al. Cefiderocol-Based versus Colistin-Based Regimens for Severe Carbapenem-Resistant *Acinetobacter baumannii* Infections: A Propensity Score-Weighted, Retrospective Cohort Study during the First Two Years of the COVID-19 Pandemic. *Microorganisms* **2023**, *11*, 984. [CrossRef]
39. Bavaro, D.F.; Papagni, R.; Belati, A.; Diella, L.; De Luca, A.; Brindicci, G.; De Gennaro, N.; Di Gennaro, F.; Romanelli, F.; Stolfa, S.; et al. Cefiderocol Versus Colistin for the Treatment of Carbapenem-Resistant *Acinetobacter baumannii* Complex Bloodstream Infections: A Retrospective, Propensity-Score Adjusted, Monocentric Cohort Study. *Infect. Dis. Ther.* **2023**, *12*, 2147–2163. [CrossRef]
40. Rando, E.; Cutuli, S.L.; Sangiorgi, F., Tanzarella, E.S.; Giovannenze, F.; De Angelis, G.; Murri, R.; Antonelli, M.; Fantoni, M.; De Pascale, G. Cefiderocol-containing regimens for the treatment of carbapenem-resistant *A. baumannii* ventilator-associated pneumonia: A propensity-weighted cohort study. *JAC-Antimicrob. Resist.* **2023**, *5*, dlad085. [CrossRef]
41. Dalfino, L.; Stufano, M.; Bavaro, D.F.; Diella, L.; Belati, A.; Stolfa, S.; Romanelli, F.; Ronga, L.; Di Mussi, R.; Murgolo, F.; et al. Effectiveness of First-Line Therapy with Old and Novel Antibiotics in Ventilator-Associated Pneumonia Caused by Carbapenem-Resistant *Acinetobacter baumannii*: A Real Life, Prospective, Observational, Single-Center Study. *Antibiotics* **2023**, *12*, 1048. [CrossRef]
42. Russo, A.; Bruni, A.; Gullì, S.; Borrazzo, C.; Quirino, A.; Lionello, R.; Serapide, F.; Garofalo, E.; Serraino, R.; Romeo, F.; et al. Efficacy of cefiderocol- vs colistin-containing regimen for treatment of bacteraemic ventilator-associated pneumonia caused by carbapenem-resistant *Acinetobacter baumannii* in patients with COVID-19. *Int. J. Antimicrob. Agents* **2023**, *62*, 106825. [CrossRef]
43. Mezcord, V.; Escalante, J.; Nishimura, B.; Traglia, G.M.; Sharma, R.; Vallé, Q.; Tuttobene, M.R.; Subils, T.; Marin, I.; Pasteran, F.; et al. Induced Heteroresistance in Carbapenem-Resistant *Acinetobacter baumannii* (CRAB) via Exposure to Human Pleural Fluid (HPF) and Its Impact on Cefiderocol Susceptibility. *Int. J. Mol. Sci.* **2023**, *24*, 11752. [CrossRef]

44. Katsube, T.; Saisho, Y.; Shimada, J.; Furuie, H. Intrapulmonary pharmacokinetics of cefiderocol, a novel siderophore cephalosporin, in healthy adult subjects. *J. Antimicrob. Chemother.* **2019**, *74*, 1971–1974. [CrossRef]
45. Wise, M.G.; Karlowsky, J.A.; Hackel, M.A.; Takemura, M.; Yamano, Y.; Echols, R.; Sahm, D.F. In Vitro Activity of Cefiderocol Against Meropenem-Nonsusceptible Gram-Negative Bacilli with Defined β-Lactamase Carriage: SIDERO-WT Surveillance Studies, 2014–2019. *Microb. Drug Resist.* **2023**, *29*, 360–370. [CrossRef]
46. Takemura, M.; Wise, M.G.; Hackel, M.A.; Sahm, D.F.; Yamano, Y. In vitro activity of cefiderocol against MBL-producing Gram-negative bacteria collected in North America and Europe in five consecutive annual multinational SIDERO-WT surveillance studies (2014–2019). *J. Antimicrob. Chemother.* **2023**, *78*, 2019–2027. [CrossRef]
47. Shortridge, D.; Streit, J.M.; Mendes, R.; Castanheira, M. In Vitro Activity of Cefiderocol against U.S. and European Gram-Negative Clinical Isolates Collected in 2020 as Part of the SENTRY Antimicrobial Surveillance Program. *Microbiol. Spectr.* **2022**, *10*, e0271221. [CrossRef] [PubMed]
48. Alzayer, M.; Alghoribi, M.F.; Alalwan, B.; Alreheli, A.; Aljohani, S.; Bosaeed, M.; Doumith, M. In vitro activity of cefiderocol against clinically important carbapenem non-susceptible Gram-negative bacteria from Saudi Arabia. *J. Glob. Antimicrob. Resist.* **2023**, *32*, 176–180. [CrossRef] [PubMed]
49. Galani, I.; Papoutsaki, V.; Karaiskos, I.; Moustakas, N.; Galani, L.; Maraki, S.; Mavromanolaki, V.E.; Legga, O.; Fountoulis, K.; Platsouka, E.D.; et al. In vitro activities of omadacycline, eravacycline, cefiderocol, apramycin, and comparator antibiotics against *Acinetobacter baumannii* causing bloodstream infections in Greece, 2020–2021: A multicenter study. *Eur. J. Clin. Microbiol. Infect. Dis.* **2023**, *42*, 843–852. [CrossRef] [PubMed]
50. Lupia, T.; Corcione, S.; Shbaklo, N.; Montrucchio, G.; De Benedetto, I.; Fornari, V.; Bosio, R.; Rizzello, B.; Mornese Pinna, S.; Brazzi, L.; et al. Meropenem/Vaborbactam and Cefiderocol as Combination or Monotherapy to Treat Multi-Drug Resistant Gram-Negative Infections: A Regional Cross-Sectional Survey from Piedmont Infectious Disease Unit Network (PIDUN). *J. Funct. Biomater.* **2022**, *13*, 174. [CrossRef]
51. Wright, H.; Harris, P.N.A.; Chatfield, M.D.; Lye, D.; Henderson, A.; Harris-Brown, T.; Donaldson, A.; Paterson, D.L. Investigator-Driven Randomised Controlled Trial of Cefiderocol versus Standard Therapy for Healthcare-Associated and Hospital-Acquired Gram-negative Bloodstream Infection: Study protocol (the GAME CHANGER trial): Study protocol for an open-label, randomised controlled trial. *Trials* **2021**, *22*, 889.
52. Liu, J.; Shu, Y.; Zhu, F.; Feng, B.; Zhang, Z.; Liu, L.; Wang, G. Comparative efficacy and safety of combination therapy with high-dose sulbactam or colistin with additional antibacterial agents for multiple drug-resistant and extensively drug-resistant *Acinetobacter baumannii* infections: A systematic review and network meta-analysis. *J. Glob. Antimicrob. Resist.* **2021**, *24*, 136–147.
53. Levin, A.S.; Barone, A.A.; Penço, J.; Santos, M.V.; Marinho, I.S.; Arruda, E.A.; Manrique, E.I.; Costa, S.F. Intravenous colistin as therapy for nosocomial infections caused by multidrug-resistant *Pseudomonas aeruginosa* and *Acinetobacter baumannii*. *Clin. Infect. Dis.* **1999**, *28*, 1008–1011. [CrossRef]
54. Markou, N.; Apostolakos, H.; Koumoudiou, C.; Athanasiou, M.; Koutsoukou, A.; Alamanos, I.; Gregorakos, L. Intravenous colistin in the treatment of sepsis from multiresistant Gram-negative bacilli in critically ill patients. *Crit. Care* **2003**, *7*, R78–R83. [CrossRef]
55. Garnacho-Montero, J.; Ortiz-Leyba, C.; Jiménez-Jiménez, F.J.; Barrero-Almodóvar, A.E.; García-Garmendia, J.L.; Bernabeu-WittelI, M.; Gallego-Lara, S.L.; Madrazo-Osuna, J. Treatment of multidrug-resistant *Acinetobacter baumannii* ventilator-associated pneumonia (VAP) with intravenous colistin: A comparison with imipenem-susceptible VAP. *Clin. Infect. Dis.* **2003**, *36*, 1111–1118. [CrossRef]
56. Reina, R.; Estenssoro, E.; Sáenz, G.; Canales, H.S.; Gonzalvo, R.; Vidal, G.; Martins, G.; Das Neves, A.; Santander, O.; Ramos, C. Safety and efficacy of colistin in *Acinetobacter* and *Pseudomonas* infections: A prospective cohort study. *Intensive Care Med.* **2005**, *31*, 1058–1065. [CrossRef] [PubMed]
57. Falagas, M.E.; Rafailidis, P.I.; Kasiakou, S.K.; Hatzopoulou, P.; Michalopoulos, A. Effectiveness and nephrotoxicity of colistin monotherapy vs. colistin-meropenem combination therapy for multidrug-resistant Gram-negative bacterial infections. *Clin. Microbiol. Infect.* **2006**, *12*, 1227–1230. [CrossRef] [PubMed]
58. Kallel, H.; Hergafi, L.; Bahloul, M.; Hakim, A.; Dammak, H.; Chelly, H.; Hamida, C.B.; Chaari, A.; Rekik, N.; Bouaziz, M. Safety and efficacy of colistin compared with imipenem in the treatment of ventilator-associated pneumonia: A matched case-control study. *Intensive Care Med.* **2007**, *33*, 1162–1167. [CrossRef] [PubMed]
59. Koomanachai, P.; Tiengrim, S.; Kiratisin, P.; Thamlikitkul, V. Efficacy and safety of colistin (colistimethate sodium) for therapy of infections caused by multidrug-resistant *Pseudomonas aeruginosa* and *Acinetobacter baumannii* in Siriraj Hospital, Bangkok, Thailand. *Int. J. Infect. Dis.* **2007**, *11*, 402–406. [CrossRef] [PubMed]
60. Falagas, M.E.; Rafailidis, P.I.; Ioannidou, E.; Alexiou, V.G.; Matthaiou, D.K.; Karageorgopoulos, D.E.; Kapaskelis, A.; Nikita, D.; Michalopoulos, A. Colistin therapy for microbiologically documented multidrug-resistant Gram-negative bacterial infections: A retrospective cohort study of 258 patients. *Int. J. Antimicrob. Agents* **2010**, *35*, 194–199. [CrossRef]
61. Durante-Mangoni, E.; Signoriello, G.; Andini, R.; Mattei, A.; De Cristoforo, M.; Murino, P.; Bassetti, M.; Malacarne, P.; Petrosillo, N.; Galdieri, N.; et al. Colistin and rifampicin compared with colistin alone for the treatment of serious infections due to extensively drug-resistant *Acinetobacter baumannii*: A multicenter, randomized clinical trial. *Clin. Infect. Dis.* **2013**, *57*, 349–358. [CrossRef] [PubMed]

62. Batirel, A.; Balkan, I.I.; Karabay, O.; Agalar, C.; Akalin, S.; Alici, O.; Alp, E.; Altay, F.A.; Altin, N.; Arslan, F.; et al. Comparison of colistin-carbapenem, colistin-sulbactam, and colistin plus other antibacterial agents for the treatment of extremely drug-resistant *Acinetobacter baumannii* bloodstream infections. *Eur. J. Clin. Microbiol. Infect. Dis.* **2014**, *33*, 1311–1322. [CrossRef] [PubMed]
63. Kalin, G.; Alp, E.; Akin, A.; Coskun, R.; Doganay, M. Comparison of colistin and colistin/sulbactam for the treatment of multidrug resistant *Acinetobacter baumannii* ventilator-associated pneumonia. *Infection* **2014**, *42*, 37–42. [CrossRef] [PubMed]
64. López-Cortés, L.E.; Cisneros, J.M.; Fernández-Cuenca, F.; Bou, G.; Tomás, M.; Garnacho-Montero, J.; Pascual, A.; Martínez-Martínez, L.; Vila, J.; Pachón, J.; et al. Monotherapy versus combination therapy for sepsis due to multidrug-resistant *Acinetobacter baumannii*: Analysis of a multicentre prospective cohort. *J. Antimicrob. Chemother.* **2014**, *69*, 3119–3126. [CrossRef] [PubMed]
65. Cheng, A.; Chuang, Y.C.; Sun, H.Y.; Sheng, W.H.; Yang, C.J.; Liao, C.H.; Hsueh, P.R.; Yang, J.L.; Shen, N.J.; Wang, J.T.; et al. Excess Mortality Associated with Colistin-Tigecycline Compared with Colistin-Carbapenem Combination Therapy for Extensively Drug-Resistant *Acinetobacter baumannii* Bacteremia: A Multicenter Prospective Observational Study. *Crit. Care Med.* **2015**, *43*, 1194–1204. [CrossRef] [PubMed]
66. Yilmaz, G.R.; Guven, T.; Guner, R.; Kocak Tufan, Z.; Izdes, S.; Tasyaran, M.A.; Acikgoz, Z.C. Colistin alone or combined with sulbactam or carbapenem against *A. baumannii* in ventilator-associated pneumonia. *J. Infect. Dev. Ctries.* **2015**, *9*, 476–485. [CrossRef]
67. Balkan, I.I.; Batirel, A.; Karabay, O.; Agalar, C.; Akalin, S.; Alici, O.; Alp, E.; Altay, F.A.; Altin, N.; Arslan, F.; et al. Comparison of colistin monotherapy and non-colistin combinations in the treatment of multi-drug resistant *Acinetobacter* spp. bloodstream infections: A multicenter retrospective analysis. *Indian. J. Pharmacol.* **2015**, *47*, 95–100. [CrossRef]
68. Amat, T.; Gutiérrez-Pizarraya, A.; Machuca, I.; Gracia-Ahufinger, I.; Pérez-Nadales, E.; Torre-Giménez, Á.; Garnacho-Montero, J.; Cisneros, J.M.; Torre-Cisneros, J. The combined use of tigecycline with high-dose colistin might not be associated with higher survival in critically ill patients with bacteraemia due to carbapenem-resistant *Acinetobacter baumannii. Clin. Microbiol. Infect.* **2018**, *24*, 630–634. [CrossRef]
69. Makris, D.; Petinaki, E.; Tsolaki, V.; Manoulakas, E.; Mantzarlis, K.; Apostolopoulou, O.; Sfyras, D.; Zakynthinos, E. Colistin versus Colistin Combined with Ampicillin-Sulbactam for Multiresistant *Acinetobacter baumannii* Ventilator-associated Pneumonia Treatment: An Open-label Prospective Study. *Indian J. Crit. Care Med.* **2018**, *22*, 67–77. [CrossRef]
70. Park, S.Y.; Si, H.J.; Eom, J.S.; Lee, J.S. Survival of carbapenem-resistant *Acinetobacter baumannii* bacteremia: Colistin monotherapy versus colistin plus meropenem. *J. Int. Med. Res.* **2019**, *47*, 5977–5985. [CrossRef]
71. Chen, C.Y.; Yang, K.Y.; Peng, C.K.; Sheu, C.C.; Chan, M.C.; Feng, J.Y.; Wang, S.H.; Chen, C.M.; Zheng, Z.R.; Liang, S.J.; et al. Clinical outcome of nosocomial pneumonia caused by Carbapenem-resistant gram-negative bacteria in critically ill patients: A multicenter retrospective observational study. *Sci. Rep.* **2022**, *12*, 7501. [CrossRef]
72. Kaye, K.S.; Marchaim, D.; Thamlikitkul, V.; Carmeli, Y.; Chiu, C.H.; Daikos, G.; Dhar, S.; Durante-Mangoni, E.; Gikas, A.; Kotanidou, A.; et al. Colistin Monotherapy versus Combination Therapy for Carbapenem-Resistant Organisms. *NEJM Evid.* **2023**, *2*, EVIDoa2200131. [CrossRef] [PubMed]
73. Paul, M.; Daikos, G.L.; Durante-Mangoni, E.; Yahav, D.; Carmeli, Y.; Benattar, Y.D.; Skiada, A.; Andini, R.; Eliakim-Raz, N.; Nutman, A. Colistin alone versus colistin plus meropenem for treatment of severe infections caused by carbapenem-resistant Gram-negative bacteria: An open-label, randomised controlled trial. *Lancet Infect. Dis.* **2018**, *18*, 391–400. [CrossRef]
74. Vardakas, K.Z.; Voulgaris, G.L.; Samonis, G.; Falagas, M.E. Inhaled colistin monotherapy for respiratory tract infections in adults without cystic fibrosis: A systematic review and meta-analysis. *Int. J. Antimicrob. Agents* **2018**, *51*, 1–9. [CrossRef] [PubMed]
75. Zhang, X.; Cui, X.; Jiang, M.; Huang, S.; Yang, M. Nebulized colistin as the adjunctive treatment for ventilator-associated pneumonia: A systematic review and meta-analysis. *J. Crit. Care* **2023**, *77*, 154315. [CrossRef] [PubMed]
76. Mohammed, N.; Savardekar, A.R.; Patra, D.P.; Narayan, V.; Nanda, A. The 21st-century challenge to neurocritical care: The rise of the superbug *Acinetobacter baumannii*. A meta-analysis of the role of intrathecal or intraventricular antimicrobial therapy in reduction of mortality. *Neurosurg. Focus.* **2017**, *43*, E8. [CrossRef] [PubMed]
77. Ng, T.M.; Teng, C.B.; Lye, D.C.; Apisarnthanarak, A. A multicenter case-case control study for risk factors and outcomes of extensively drug-resistant *Acinetobacter baumannii* bacteremia. *Infect. Control Hosp. Epidemiol.* **2014**, *35*, 49–55. [CrossRef] [PubMed]
78. Phe, K.; Lee, Y.; McDaneld, P.M.; Prasad, N.; Yin, T.; Figueroa, D.A.; Musick, W.L.; Cottreau, J.M.; Hu, M.; Tam, V.H. In vitro assessment and multicenter cohort study of comparative nephrotoxicity rates associated with colistimethate versus polymyxin B therapy. *Antimicrob. Agents Chemother.* **2014**, *58*, 2740–2746. [CrossRef] [PubMed]
79. Crusio, R.; Rao, S.; Changawala, N.; Paul, V.; Tiu, C.; van Ginkel, J.; Chapnick, E.; Kupfer, Y. Epidemiology and outcome of infections with carbapenem-resistant Gram-negative bacteria treated with polymyxin B-based combination therapy. *Scand. J. Infect. Dis.* **2014**, *46*, 1–8. [CrossRef]
80. Nelson, B.C.; Eiras, D.P.; Gomez-Simmonds, A.; Loo, A.S.; Satlin, M.J.; Jenkins, S.G.; Whittier, S.; Calfee, D.P.; Furuya, E.Y.; Kubin, C.J. Clinical outcomes associated with polymyxin B dose in patients with bloodstream infections due to carbapenem-resistant Gram-negative rods. *Antimicrob. Agents Chemother.* **2015**, *59*, 7000–7006. [CrossRef]
81. Rigatto, M.H.; Behle, T.F.; Falci, D.R.; Freitas, T.; Lopes, N.T.; Nunes, M.; Costa, L.W.; Zavascki, A.P. Risk factors for acute kidney injury (AKI) in patients treated with polymyxin B and influence of AKI on mortality: A multicentre prospective cohort study. *J. Antimicrob. Chemother.* **2015**, *70*, 1552–1557. [CrossRef] [PubMed]

82. Rigatto, M.H.; Vieira, F.J.; Antochevis, L.C.; Behle, T.F.; Lopes, N.T.; Zavascki, A.P. Polymyxin B in Combination with Antimicrobials Lacking In Vitro Activity versus Polymyxin B in Monotherapy in Critically Ill Patients with *Acinetobacter baumannii* or *Pseudomonas aeruginosa* Infections. *Antimicrob. Agents Chemother.* **2015**, *59*, 6575–6580. [CrossRef]
83. Rigatto, M.H.; Oliveira, M.S.; Perdigão-Neto, L.V.; Levin, A.S.; Carrilho, C.M.; Tanita, M.T.; Tuon, F.F.; Cardoso, D.E.; Lopes, N.T.; Falci, D.R.; et al. Multicenter Prospective Cohort Study of Renal Failure in Patients Treated with Colistin versus Polymyxin B. *Antimicrob. Agents Chemother.* **2016**, *60*, 2443–2449. [CrossRef]
84. Cai, B.; Cai, Y.; Liew, Y.X.; Chua, N.G.; Teo, J.Q.M.; Lim, T.P.; Kurup, A.; Ee, P.L.; Tan, T.T.; Lee, W.; et al. Clinical efficacy of polymyxin monotherapy versus nonvalidated polymyxin combination therapy versus validated polymyxin combination therapy in extensively drug-resistant Gram-negative bacillus infections. *Antimicrob. Agents Chemother.* **2016**, *60*, 4013–4022. [CrossRef]
85. Mattos, K.P.H.; Gouvêa, I.R.; Quintanilha, J.C.F.; Cursino, M.A.; Vasconcelos, P.E.N.S.; Moriel, P. Polymyxin B clinical outcomes: A prospective study of patients undergoing intravenous treatment. *J. Clin. Pharm. Ther.* **2019**, *44*, 415–419. [CrossRef] [PubMed]
86. Pan, S.; Huang, X.; Wang, Y.; Li, L.; Zhao, C.; Yao, Z.; Cui, W.; Zhang, G. Efficacy of intravenous plus intrathecal/intracerebral ventricle injection of polymyxin B for post-neurosurgical intracranial infections due to MDR/XDR *Acinectobacter baumannii*: A retrospective cohort study. *Antimicrob. Resist. Infect. Control* **2018**, *7*, 8. [CrossRef] [PubMed]
87. Simon, V.; Viswam, A.; Alexander, P.S.; James, E.; Sudhindran, S. Colistin versus polymyxin B: A pragmatic assessment of renal and neurological adverse effects and effectiveness in multidrug-resistant Gram-negative bacterial infections. *Indian J. Pharmacol.* **2023**, *55*, 229–236.
88. Liu, S.; Wu, Y.; Qi, S.; Shao, H.; Feng, M.; Xing, L.; Liu, H.; Gao, Y.; Zhu, Z.; Zhang, S.; et al. Polymyxin B therapy based on therapeutic drug monitoring in carbapenem-resistant organisms sepsis: The PMB-CROS randomized clinical trial. *Crit. Care* **2023**, *27*, 232. [CrossRef]
89. Garcia, R.C.L.; Rodrigues, R.D.; Garcia, E.C.L.; Rigatto, M.H. Comparison between Colistin and Polymyxin B in the Treatment of Bloodstream Infections Caused by Carbapenem-resistant *Pseudomonas aeruginosa* and *Acinetobacter baumannii-calcoaceticus* Complex. *Antibiotics* **2023**, *12*, 1317. [CrossRef]
90. Chang, K.; Wang, H.; Zhao, J.; Yang, X.; Wu, B.; Sun, W.; Huang, M.; Cheng, Z.; Chen, H.; Song, Y.; et al. Polymyxin B/Tigecycline Combination vs. Polymyxin B or Tigecycline Alone for the Treatment of Hospital-Acquired Pneumonia Caused by Carbapenem-Resistant *Enterobacteriaceae* or Carbapenem-Resistant *Acinetobacter baumannii*. *Front. Med.* **2022**, *9*, 772372. [CrossRef]
91. Zha, L.; Zhang, X.; Cheng, Y.; Xu, Q.; Liu, L.; Chen, S.; Lu, Z.; Guo, J.; Tefsen, B. Intravenous Polymyxin B as Adjunctive Therapy to High-Dose Tigecycline for the Treatment of Nosocomial Pneumonia Due to Carbapenem-Resistant *Acinetobacter baumannii* and *Klebsiella pneumoniae*: A Propensity Score-Matched Cohort Study. *Antibiotics* **2023**, *12*, 273. [CrossRef] [PubMed]
92. Qiao, L.; Zuo, W.; Yang, Y.; Liu, X.; Wang, Q.; Yu, J.; Wu, J.; Xu, T.; Jiang, J.; Zhang, B.; et al. Clinical outcomes and safety of intravenous polymyxin B-based treatment in critically ill patients with carbapenem-resistant *Acinetobacter baumannii* nosocomial pneumonia. *Int. J. Antimicrob. Agents* **2023**, *62*, 106880. [CrossRef] [PubMed]
93. Yu, L.L.; Shi, X.P.; Huang, J.F.; Gong, Y.; Cui, C.X.; Wang, T. A retrospective observational study of the treatment with polymyxin B for liver transplantation recipients infected by multidrug-resistant gram-negative bacteria. *J. Clin. Pharm. Ther.* **2022**, *47*, 1563–1569. [CrossRef] [PubMed]
94. Ngamprasertchai, T.; Boonyasiri, A.; Charoenpong, L.; Nimitvilai, S.; Lorchirachoonkul, N.; Wattanamongkonsil, L.; Thamlikitkul, V. Effectiveness and safety of polymyxin B for the treatment of infections caused by extensively drug-resistant Gram-negative bacteria in Thailand. *Infect. Drug Resist.* **2018**, *11*, 1219–1224. [CrossRef] [PubMed]
95. Quintanilha, J.C.F.; Duarte, N.D.C.; Lloret, G.R.; Visacri, M.B.; Mattos, K.P.H.; Dragosavac, D.; Falcão, A.L.E.; Moriel, P. Colistin and polymyxin B for treatment of nosocomial infections in intensive care unit patients: Pharmacoeconomic analysis. *Int. J. Clin. Pharm.* **2019**, *41*, 74–80. [CrossRef] [PubMed]
96. Baradaran, S.; Black, D.J.; Keyloun, K.R.; Hansen, R.N.; Gillard, P.J.; Devine, B. The Impact of Acute Kidney Injury on the Risk of Mortality and Health Care Utilization Among Patients Treated with Polymyxins for Severe Gram-Negative Infections. *Open Forum Infect. Dis.* **2018**, *5*, ofy191. [CrossRef]
97. Wang, J.; Shah, B.K.; Zhao, J.; Xiong, J.; Wang, C.; Xie, S. Comparative study of polymyxin B and colistin sulfate in the treatment of severe comorbid patients infected with CR-GNB. *BMC Infect. Dis.* **2023**, *23*, 351. [CrossRef]
98. Ismail, B.; Shafei, M.N.; Harun, A.; Ali, S.; Omar, M.; Deris, Z.Z. Predictors of polymyxin B treatment failure in Gram-negative healthcare-associated infections among critically ill patients. *J. Microbiol. Immunol. Infect.* **2018**, *51*, 763–769. [CrossRef]
99. Pereira, G.H.; Muller, P.R.; Levin, A.S. Salvage treatment of pneumonia and initial treatment of tracheobronchitis caused by multidrug-resistant Gram-negative bacilli with inhaled polymyxin B. *Diagn. Microbiol. Infect. Dis.* **2007**, *58*, 235–240. [CrossRef]
100. Okoduwa, A.; Ahmed, N.; Guo, Y.; Scipione, M.R.; Papadopoulos, J.; Eiras, D.P.; Dubrovskaya, Y. Nephrotoxicity Associated with Intravenous Polymyxin B Once- versus Twice-Daily Dosing Regimen. *Antimicrob. Agents Chemother.* **2018**, *62*, e00025-18. [CrossRef]
101. John, J.F.; Falci, D.R.; Rigatto, M.H.; Oliveira, R.D.; Kremer, T.G.; Zavascki, A.P. Severe Infusion-Related Adverse Events and Renal Failure in Patients Receiving High-Dose Intravenous Polymyxin B. *Antimicrob. Agents Chemother.* **2017**, *62*, e01617-17. [CrossRef]
102. Oliota, A.F.; Penteado, S.T.; Tonin, F.S.; Fernandez-Llimos, F.; Sanches, A.C. Nephrotoxicity prevalence in patients treated with polymyxins: A systematic review with meta-analysis of observational studies. *Diagn. Microbiol. Infect. Dis.* **2019**, *94*, 41–49. [CrossRef] [PubMed]

103. Seifert, H.; Blondeau, J.; Lucaßen, K.; Utt, E.A. Global update on the in vitro activity of tigecycline and comparators against isolates of *Acinetobacter baumannii* and rates of resistant phenotypes (2016–2018). *J. Glob. Antimicrob. Resist.* **2022**, *31*, 82–89. [CrossRef]
104. Hua, X.; He, J.; Wang, J.; Zhang, L.; Zhang, L.; Xu, Q.; Shi, K.; Leptihn, S.; Shi, Y.; Fu, X.; et al. Novel tigecycline resistance mechanisms in *Acinetobacter baumannii* mediated by mutations in *adeS*, *rpoB* and *rrf*. *Emerg. Microbes Infect.* **2021**, *10*, 1404–1417. [CrossRef] [PubMed]
105. Barbour, A.; Schmidt, S.; Ma, B.; Schiefelbein, L.; Rand, K.H.; Burkhardt, O.; Derendorf, H. Clinical pharmacokinetics and pharmacodynamics of tigecycline. *Clin. Pharmacokinet.* **2009**, *48*, 575–584. [CrossRef] [PubMed]
106. Gotfried, M.H.; Horn, K.; Garrity-Ryan, L.; Villano, S.; Tzanis, E.; Chitra, S.; Manley, A.; Tanaka, S.K.; Rodvold, K.A. Comparison of Omadacycline and Tigecycline Pharmacokinetics in the Plasma, Epithelial Lining Fluid, and Alveolar Cells of Healthy Adult Subjects. *Antimicrob. Agents Chemother.* **2017**, *61*, e01135-17. [CrossRef] [PubMed]
107. De Pascale, G.; Lisi, L.; Ciotti, G.M.P.; Vallecoccia, M.S.; Cutuli, S.L.; Cascarano, L.; Gelormini, C.; Bello, G.; Montini, L.; Carelli, S.; et al. Pharmacokinetics of high-dose tigecycline in critically ill patients with severe infections. *Ann. Intensive Care* **2020**, *10*, 94. [CrossRef] [PubMed]
108. Sodeifian, F.; Zangiabadian, M.; Arabpour, E.; Kian, N.; Yazarlou, F.; Goudarzi, M.; Centis, R.; Seghatoleslami, Z.S.; Kameh, M.C.; Danaei, B.; et al. Tigecycline-Containing Regimens and Multi Drug-Resistant *Acinetobacter baumannii*: A Systematic Review and Meta-Analysis. *Microb. Drug Resist.* **2023**, *29*, 344–359. [CrossRef]
109. Gong, J.; Su, D.; Shang, J.; Yu, H.; Du, G.; Lin, Y.; Sun, Z.; Liu, G. Efficacy and safety of high-dose tigecycline for the treatment of infectious diseases: A meta-analysis. *Medicine* **2019**, *98*, e17091. [CrossRef]
110. Chuang, Y.C.; Cheng, C.Y.; Sheng, W.H.; Sun, H.Y.; Wang, J.T.; Chen, Y.C.; Chang, S.C. Effectiveness of tigecycline-based versus colistin- based therapy for treatment of pneumonia caused by multidrug-resistant *Acinetobacter baumannii* in a critical setting: A matched cohort analysis. *BMC Infect. Dis.* **2014**, *14*, 102. [CrossRef]
111. Kwon, S.H.; Ahn, H.L.; Han, O.Y.; La, H.O. Efficacy and safety profile comparison of colistin and tigecycline on the extensively drug resistant *Acinetobacter baumannii*. *Biol. Pharm. Bull.* **2014**, *37*, 340–346. [CrossRef]
112. Chaari, A.; Pham, T.; Mnif, B.; Chtara, K.; Medhioub, F.; Baccouche, N.; Bahloul, M.; Hammami, A.; Bouaziz, M. Colistin-tigecycline versus colistin-imipenem-cilastatin combinations for the treatment of *Acinetobacter baumannii* ventilator-acquired pneumonia: A prognosis study. *Intensive Care Med.* **2015**, *41*, 2018–2019. [CrossRef] [PubMed]
113. Kim, W.Y.; Moon, J.Y.; Huh, J.W.; Choi, S.H.; Lim, C.M.; Koh, Y.; Chong, Y.P.; Hong, S.B. Comparable Efficacy of Tigecycline versus Colistin Therapy for Multidrug-Resistant and Extensively Drug-Resistant *Acinetobacter baumannii* Pneumonia in Critically Ill Patients. *PLoS ONE* **2016**, *11*, e0150642. [CrossRef] [PubMed]
114. Ye, J.J.; Lin, H.S.; Yeh, C.F.; Wu, Y.M.; Huang, P.Y.; Yang, C.C.; Huang, C.T.; Lee, M.H. Tigecycline-based versus sulbactam-based treatment for pneumonia involving multidrug-resistant *Acinetobacter calcoaceticus-Acinetobacter baumannii* complex. *BMC Infect. Dis.* **2016**, *16*, 374. [CrossRef]
115. Jean, S.S.; Hsieh, T.C.; Lee, W.S.; Hsueh, P.R.; Hsu, C.W.; Lam, C. Treatment outcomes of patients with non-bacteremic pneumonia caused by extensively drug-resistant *Acinetobacter calcoaceticus-Acinetobacter baumannii* complex isolates: Is there any benefit of adding tigecycline to aerosolized colistimethate sodium? *Medicine* **2018**, *97*, e12278. [CrossRef]
116. Niu, T.; Luo, Q.; Li, Y.; Zhou, Y.; Yu, W.; Xiao, Y. Comparison of Tigecycline or Cefoperazone/Sulbactam therapy for bloodstream infection due to Carbapenem-resistant *Acinetobacter baumannii*. *Antimicrob. Resist. Infect. Control* **2019**, *8*, 52. [CrossRef] [PubMed]
117. Hoban, D.J.; Reinert, R.R.; Bouchillon, S.K.; Dowzicky, M.J. Global in vitro activity of tigecycline and comparator agents: Tigecycline Evaluation and Surveillance Trial 2004–2013. *Ann. Clin. Microbiol. Antimicrob.* **2015**, *14*, 27. [CrossRef] [PubMed]
118. Lodise, T.P.; Van Wart, S.; Sund, Z.M.; Bressler, A.M.; Khan, A.; Makley, A.T.; Hamad, Y.; Salata, R.A.; Silveira, F.P.; Sims, M.D.; et al. Pharmacokinetic and Pharmacodynamic Profiling of Minocycline for Injection following a Single Infusion in Critically Ill Adults in a Phase IV Open-Label Multicenter Study (ACUMIN). *Antimicrob. Agents Chemother.* **2021**, *65*, e01809-20. [CrossRef] [PubMed]
119. Fragkou, P.C.; Poulakou, G.; Blizou, A.; Blizou, M.; Rapti, V.; Karageorgopoulos, D.E.; Koulenti, D.; Papadopoulos, A.; Matthaiou, D.K.; Tsiodras, S. The Role of Minocycline in the Treatment of Nosocomial Infections Caused by Multidrug, Extensively Drug and Pandrug Resistant *Acinetobacter baumannii*: A Systematic Review of Clinical Evidence. *Microorganisms* **2019**, *7*, 159. [CrossRef]
120. Pogue, J.M.; Neelakanta, A.; Mynatt, R.P.; Sharma, S.; Lephart, P.; Kaye, K.S. Carbapenem-resistance in gram-negative bacilli and intravenous minocycline: An antimicrobial stewardship approach at the Detroit Medical Center. *Clin. Infect. Dis.* **2014**, *59*, S388–S393. [CrossRef]
121. Goff, D.A.; Bauer, K.A.; Mangino, J.E. Bad bugs need old drugs: A stewardship program's evaluation of minocycline for multidrug-resistant *Acinetobacter baumannii* infections. *Clin. Infect. Dis.* **2014**, *59*, S381–S387. [CrossRef]
122. Wang, L.; Zou, L.; Li, H.; Li, T. Preliminary Analysis on the Treatment of Ventilator-Associated Pneumonia Caused by Pandrug-Resistant *Acinetobacter baumannii*. *Zhongguo Yi XueKeXue Yuan Xue Bao* **2014**, *36*, 185–188.
123. Bremmer, D.N.; Bauer, K.A.; Pouch, S.M.; Thomas, K.; Smith, D.; Goff, D.A.; Pancholi, P.; Balada-Llasat, J.M. Correlation of Checkerboard Synergy Testing with Time-Kill Analysis and Clinical Outcomes of Extensively Drug-Resistant *Acinetobacter baumannii* Respiratory Infections. *Antimicrob. Agents Chemother.* **2016**, *60*, 6892–6895. [CrossRef]

124. Bishburg, E.; Shah, M.; Chan, T. Use of Intravenous Minocycline for the Treatment of Methicillin-Resistant Staphylococcus aureus (MRSA) and Resistant Gram-Negati132-135ve Organisms: Experience in a Tertiary Care Hospital. *Infect. Dis. Clin. Pract.* **2014**, *22*, 26–31. [CrossRef]
125. Lomovskaya, O.; Sun, D.; Rubio-Aparicio, D.; Nelson, K.J.; Thamlikitkul, V.; Dudley, M.N.; Redell, M.A. Absence of TetB identifies minocycline-susceptible isolates of *Acinetobacter baumannii*. *Int. J. Antimicrob. Agents* **2018**, *52*, 404–406. [CrossRef]
126. Raz-Pasteur, A.; Liron, Y.; Amir-Ronen, R.; Abdelgani, S.; Ohanyan, A.; Geffen, Y.; Paul, M. Trimethoprim-sulfamethoxazole vs. colistin or ampicillin-sulbactam for the treatment of carbapenem-resistant *Acinetobacter baumannii*: A retrospective matched cohort study. *J. Glob. Antimicrob. Resist.* **2019**, *17*, 168–172. [CrossRef]
127. Sirijatuphat, R.; Thamlikitkul, V. Preliminary study of colistin versus colistin plus fosfomycin for treatment of carbapenem-resistant *Acinetobacter baumannii* infections. *Antimicrob. Agents Chemother.* **2014**, *58*, 5598–5601. [CrossRef] [PubMed]
128. Russo, A.; Bassetti, M.; Bellelli, V.; Bianchi, L.; Marincola Cattaneo, F.; Mazzocchetti, S.; Paciacconi, E.; Cottini, F.; Schiattarella, A.; Tufaro, G.; et al. Efficacy of a Fosfomycin-Containing Regimen for Treatment of Severe Pneumonia Caused by Multidrug-Resistant *Acinetobacter baumannii*: A Prospective, Observational Study. *Infect. Dis. Ther.* **2021**, *10*, 187–200. [CrossRef] [PubMed]
129. Park, H.J.; Cho, J.H.; Kim, H.J.; Han, S.H.; Jeong, S.H.; Byun, M.K. Colistin monotherapy versus colistin/rifampicin combination therapy in pneumonia caused by colistin-resistant *Acinetobacter baumannii*: A randomised controlled trial. *J. Glob. Antimicrob. Resist.* **2019**, *17*, 66–71. [CrossRef]
130. Aydemir, H.; Akduman, D.; Piskin, N.; Comert, F.; Horuz, E.; Terzi, A.; Kokturk, F.; Ornek, T.; Celebi, G. Colistin vs. the combination of colistin and rifampicin for the treatment of carbapenem-resistant *Acinetobacter baumannii* ventilator-associated pneumonia. *Epidemiol. Infect.* **2013**, *141*, 1214–1222. [CrossRef]
131. Durante-Mangoni, E.; Andini, R.; Signoriello, S.; Cavezza, G.; Murino, P.; Buono, S.; De Cristofaro, M.; Taglialatela, C.; Bassetti, M.; Malacarne, P.; et al. Acute kidney injury during colistin therapy: A prospective study in patients with extensively-drug resistant *Acinetobacter baumannii* infections. *Clin. Microbiol. Infect.* **2016**, *22*, 984–989. [CrossRef] [PubMed]
132. Tsuji, B.T.; Pogue, J.M.; Zavascki, A.P.; Paul, M.; Daikos, G.L.; Forrest, A.; Giacobbe, D.R.; Viscoli, C.; Giamarellou, H.; Karaiskos, I.; et al. International Consensus Guidelines for the Optimal Use of the Polymyxins: Endorsed by the American College of Clinical Pharmacy (ACCP), European Society of Clinical Microbiology and Infectious Diseases (ESCMID), Infectious Diseases Society of America (IDSA), International Society for Anti-infective Pharmacology (ISAP), Society of Critical Care Medicine (SCCM), and Society of Infectious Diseases Pharmacists (SIDP). *Pharmacotherapy* **2019**, *39*, 10–39. [PubMed]

antibiotics

Review

Antibiotic Treatment of Carbapenem-Resistant *Acinetobacter baumannii* Infections in View of the Newly Developed β-Lactams: A Narrative Review of the Existing Evidence

Francesca Serapide [1,†], Maurizio Guastalegname [1,†], Sara Palma Gullì [1], Rosaria Lionello [1], Andrea Bruni [2], Eugenio Garofalo [2], Federico Longhini [2], Enrico Maria Trecarichi [1,‡] and Alessandro Russo [1,*,‡]

1 Infectious and Tropical Disease Unit, Department of Medical and Surgical Sciences, 'Magna Graecia' University of Catanzaro, 88100 Catanzaro, Italy; francescaserapide@gmail.com (F.S.); sarapalma08@gmail.com (S.P.G.); rosarialionello0@gmail.com (R.L.)
2 Intensive Care Unit, Department of Medical and Surgical Sciences, 'Magna Graecia' University of Catanzaro, 88100 Catanzaro, Italy; andreabruni@unicz.it (A.B.); eugenio.garofalo@unicz.it (E.G.); flonghini@unicz.it (F.L.)
* Correspondence: a.russo@unicz.it
† These authors contributed equally to this work.
‡ These authors also contributed equally to this work.

Abstract: It is estimated that antimicrobial resistance (AMR) is responsible for nearly 5 million human deaths worldwide each year and will reach 10 million by 2050. Carbapenem-resistant *Acinetobacter baumannii* (CRAB) infections represent the fourth-leading cause of death attributable to antimicrobial resistance globally, but a standardized therapy is still lacking. Among the antibiotics under consideration, Sulbactam/durlobactam seems to be the best candidate to replace current back-bone agents. Cefiderocol could play a pivotal role within combination therapy regimens. Due to toxicity and the pharmacokinetics/pharmacodynamics (PK/PD) limitations, colistin (or polymyxin B) should be used as an alternative agent (when no other options are available). Tigecycline (or minocycline) and fosfomycin could represent suitable partners for both NBLs. Randomized clinical trials (RCTs) are needed to better evaluate the role of NBLs in CRAB infection treatment and to compare the efficacy of tigecycline and fosfomycin as partner antibiotics. Synergism should be tested between NBLs and "old" drugs (rifampicin and trimethoprim/sulfamethoxazole). Huge efforts should be made to accelerate pre-clinical and clinical studies on safer polymyxin candidates with improved lung activity, as well as on the iv rifabutin formulation. In this narrative review, we focused the antibiotic treatment of CRAB infections in view of newly developed β-lactam agents (NBLs).

Keywords: *Acinetobacter baumannii*; carbapenem resistance; sulbactam/durlobactam; cefiderocol; guidelines; real-world evidence

1. Introduction

The species belonging to *Acinetobacter baumannii-calcoaceticus* complex (ABC) are glucose-non-fermentative, aerobic Gram-negative coccobacilli [1]. The predominant predispositions to ABC infection are exposure to broad-spectrum antibiotics and disruption of anatomical barriers [2]. Among ABC, *A. baumannii* is the most common cause of human infections, responsible for a range of nosocomial infection across multiple anatomical sites, mainly ventilator-associated pneumonia (VAP) and central line-associated bloodstream infections (BSI) [3]. Peculiar findings of *A. baumannii* are the capacity to survive in unfavorable conditions, due to molecular features that promote environmental persistence (i.e., desiccation resistance, biofilm formation, and motility) and the ability to acquire or upregulate various resistance determinants (i.e., intrinsic and acquired β-lactamases, upregulation of efflux pumps, decreased outer membrane permeability, antibiotic target site modifications) [4]. Indeed, it is one of the microorganisms in the ESKAPE group, which have been identified as pathogens particularly characterized by increasing multiresistance and virulence dependent on mechanisms capable of evading the bactericidal action of antibiotics and is included in the list of critical priorities for antibiotic resistance among pathogens, along with *P. aeruginosa* (carbapenemase-resistant), *K. pneumoniae* and *Enterobacter* spp. (ESBL).

In addition, the World Health Organization (WHO) has recently included ESKAPE pathogens in the list of 12 bacteria for which new antibiotics are needed. Carbapenem resistance against *A. baumannii* varies between 30% and 80% and is commonly associated with the horizontal transfer of genes encoding oxacillinase (OXA) carbapenemases enzymes (including OXA-23 and OXA-24/40 enzymes) [5].

Carbapenem-resistant (CR)—*A. baumannii* infections are the fourth-leading cause of death attributable to antimicrobial resistance globally [6], but a standardized approach to antibiotic therapy is still lacking. Current CRAB treatment guidelines agree on the use of a combination regimen for severe infections, although there are some differences in the choice of antibiotic agents [7,8] (see Table 1).

Table 1. Differences between CRAB treatment guidelines.

	ESCMID Guidelines (April 2022)	IDSA Guidance (July 2023)
Combination antibiotic regimen	For severe and high-risk CRAB infection	For moderate–severe CRAB infection
Ampicillin/sulbactam	For patients with CRAB susceptible to sulbactam and HAP/VAP (1 g sulbactam component q6h)	Back-bone treatment for all CRAB infection (6–9 g sulbactam component daily)
Polymyxins	Either colistin or polymyxin B: for patients with CRAB resistant to sulbactam susceptible to polymyxins; in combination with one other in vitro active agent for severe, susceptible to polymyxins, CRAB infection	Polymyxin B in combination with at least one other agent for the treatment of CRAB infections (Colistin only for CRAB UTIs)
Tetracycline derivatives	High-dose tigecycline: for patients with CRAB resistant to sulbactam susceptible to tigecycline; in combination with one other in vitro active agent for severe, susceptible to tigecycline, CRAB infection	High-dose minocycline (preferred option) or high-dose tigecycline in combination with at least one other agent for the treatment of CRAB infections
Cefiderocol	Not recommended	In combination with at least one other agent for the treatment of CRAB infections refractory to other antibiotics (or when the use of other antibiotics is precluded)
Aminoglycosides	In combination with one other in vitro active agent for severe, susceptible to aminoglycosides, CRAB infection	Not recommended
High-dose extended-infusion meropenem	In combination with one other in vitro active agent for severe CRAB infections with a meropenem MIC < 8 mg/L	Not recommended

Legend. CRAB: carbapenem-resistant *Acinetobacter baumannii*; ESCMID: European Society of Clinical Microbiology and Infectious Diseases; IDSA: Infectious Diseases Society of America; HAP: health-care associated pneumonia; VAP: ventilator-associated pneumonia; UTIs: urinary tract infections. High-dose tigecycline: 200 mg as a loading dose followed by 100 mg q12h; High-dose minocycline: 200 mg q12h. High-dose extended-infusion meropenem: 2 g over 3 h infusion q8h.

The recent approval of sulbactam/durlobactam for the treatment of ABC pneumonia [9], together with the encouraging data from the real-world clinical use of cefiderocol [10,11], opens a new scenario for CRAB infections treatment.

In this narrative review, we focused the antibiotic treatment of CRAB infections in view of newly developed β-lactam agents (NBLs), presenting the available data, discussing the main hot points, and highlighting clinical questions awaiting further investigation.

2. Considerations and Available Data about New β-Lactam Agents

Beta-lactamases are classified into four groups (A, B, C, D) according to Ambler's classification and into four categories (based on their biochemical function) according to Jacoby's classification.

2.1. Sulbactam/Durlobactam

Sulbactam/durlobactam was recently approved in the U.S. for the treatment of pneumonia due to susceptible ABC [9]. Sulbactam is a class A β-lactamase inhibitor with intrinsic whole-cell activity against few bacterial species; its activity against *A. baumannii* is mediated through inhibition of the penicillin-binding protein (PBP)1a, PBP1b, and PBP3 [12]. However, sulbactam is susceptible to cleavage by several β-lactamases, such as TEM-1, Acinetobacter-derived cephalosporinase (ADC)-30, and OXAs (OXA-23, OXA-24/72, and OXA-58 families), hence its clinical utility for *A. baumannii* infections has been compromised over time [13]. Durlobactam is a novel non–β-lactam diazabicyclooctane β-lactamase inhibitor, with a broad-spectrum activity against class A, C, and D β-lactamases and PBPs, resulting in intrinsic antibacterial activity against *Enterobacterales* and restoration of β-lactam activity in (Multidrug Resistance) MDR Gram-negative pathogens [14]. In contrast to the other diazabicyclooctanes, durlobactam plays a crucial role in CRAB infection, due to the potent inhibition not only of class A β-lactamases but also of ADCs and OXAs, and the consequent ability to restore sulbactam activity. Furthermore, through PBP2 inhibition, it also showed a minimal intrinsic activity against the pathogen [14]. CRAB resistance to sulbactam/durlobactam was estimated to be 3.4%, mainly due to the substitution in the PBP3 determinant or the presence of New Delhi metallo-β-lactamase-1 (NDM-1); metallo β-lactamase (MBL)-producing strains are characterized by higher MIC values compared to other resistance mechanisms [15]. Although the current frequency of MBL-producing CRAB is relatively low [4,15], the commercialization of sulbactam/durlobactam may lead to an increase in the incidence of such strains, particularly NDM, for which *A. baumannii* has been considered a reservoir [2].

Sulbactam/durlobactam demonstrated a good intrapulmonary penetration ratio for epithelial lining fluid (ELF) to total plasma concentrations in healthy subjects, supporting the use of the combination in the treatment of pulmonary infections [16]. The ATTACK trial compared sulbactam/durlobactam (1 g/1 g over 3 h infusion q6h for 7–14 days) with colistin (2.5 mg/kg over 30 min infusion q12h for 7–14 days) for severe CRAB infection (mainly pneumonia, including VAP); both regimens were co-administered with imipenem/cilastatin as background therapy: sulbactam/durlobactam resulted non-inferior to colistin in 28-day all-cause mortality 19% (12/63) vs. 32% (20/62) presenting less adverse events (mainly nephrotoxicity, headache, nausea, and injection-site phlebitis) with respect to the competitor [17].

Co-administration of imipenem/cilastatin to provide coverage for eventual polymicrobial infections has raised questions about a possible synergic action of imipenem with sulbactam/durlobactam on CRAB, casting doubts on the real efficacy of the new β-lactam agent as monotherapy [18]. In vitro data on CRAB isolates of the ATTACK trial seem to indicate that imipenem does not provide a meaningful contribution to sulbactam–durlobactam activity against sulbactam–durlobactam-susceptible isolates [19]; however, a study conducted on 109 CRAB isolates in Greece showed that the addition of imipenem further lowered the MIC90 of sulbactam/durlobactam by one two-fold dilution [20]. Currently there are not ongoing trials on sulbactam/durlobactam [21].

The encouraging data seem to place sulbactam/durlobactam as a crucial player in CRAB infection, although two relevant clinical questions are still outstanding:

(1) What is the best antibiotic partner to use in combination with sulbactam/durlobactam in CRAB infections?
(2) Is sulbactam/durlobactam best used in combination treatment rather than monotherapy in CRAB infections?

In our opinion, studies addressing the first question should be prioritized for the following reasons: (i) the ATTACK trial tested sulbactam/durlobactam in a combination regimen, and the real role of imipenem addition has not yet been clearly elucidated; (ii) although there is no clear evidence of the superiority of a combination treatment over monotherapy in CRAB infection, currently the main guidelines agree on the use of combination treatment (at least for severe infections) [7,8]; (iii) the delay in placing in the therapy of cefiderocol in CRAB infections has been essentially due to unsatisfactory clinical results in trials testing the drug as monotherapy [22,23].

It would therefore be more prudent and appropriate, in our opinion, to prioritize the search for the best antibiotic partner of sulbactam/durlobactam and, subsequently, compare monotherapy vs. combination treatment.

2.2. Cefiderocol

Cefiderocol is a novel catechol-substituted siderophore cephalosporin, commercialized for the treatment of infections caused by CR Gram-negative bacteria (CR-GNB). Its activity is expressed through the inhibition of PBPs (primarily PBP3) and is transported into the periplasmic space mainly through the bacterial siderophore iron uptake system [24]. The unique structure of cefiderocol, along with low catalytic efficiencies of carbapenemases against this drug, confers stability against all four Ambler classes of β-lactamases [25]. The cefiderocol susceptibility rate for CRAB was estimated to be between 77.9% and 97.2% across different countries [25]. In the SIDERO surveillance program, 3.9% (204 out of 5225) of CRAB isolates had high (≥8 μg/mL) cefiderocol MICs without prior exposure to this antibiotic [26]: the β-lactamases Pseudomonas-extended resistance (PER) enzyme was detected in most of these isolates [25]. Although PER (and, to a lesser extent, NDM) contributes to increasing the cefiderocol MIC of *A. baumannii*, it does not alone confer resistance to the drug; cefiderocol resistance seems, in fact, to be mediated by the concomitant presence of other factors (such as PBP-3 modification, reduced expression of the siderophore receptor, and efflux overexpression) [27].

It is important to consider the phenomenon of heteroresistance, which is defined as a condition in which certain subpopulations of a biological sample exhibit varying degrees of phenotypic resistance, making in vitro identification of resistance difficult. In fact, some diagnostic methods may not be sensitive enough to identify these strains and may misclassify them as susceptible. However, by using appropriate antimicrobial susceptibility tests with higher inoculum, it is possible to detect even these subpopulations with intermediate or resistant minimal inhibitory concentrations.

In a study in the United States of 108 CRAB isolates, the frequency of resistant subpopulations was ≥1 in 106. In contrast, the resistance rate was 8% and the heteroresistance rate was 59% (64 isolates). In addition, the frequency of resistant subpopulations increased after exposure to cefiderocol and decreased after discontinuation of the drug. [28]. The authors of the survey hypothesized that heteroresistance could explain the discrepancy between the excellent in vitro susceptibility profile and the suboptimal clinical outcomes when used as monotherapy against CRAB [28]. However, no clinical data support this hypothesis; therefore, the real clinical impact of heteroresistance is yet unclear [29].

Cefiderocol showed an effective lung penetration in healthy subjects as well as in patients with pneumonia requiring mechanical ventilation [27]. The probability of target attainment (PTA) for 100% fT > MIC was >90% across all infection sites for pathogens with cefiderocol MICs of ≤4 μg/mL; adequate plasma exposure can be achieved at the drug recommended dosing regimen (2 g over 3 h infusion q8h) for the infected patients [30].

Despite the in vitro efficacy and PK/PD characteristics, clinical results on cefiderocol monotherapy were unsatisfactory [22,23]. The CREDIBLE-CR trial compared cefiderocol with the best available treatment (BAT) for CR-GNB: in the subgroups of patients with CRAB, all-cause mortality at day 28 (± 3 days) after the end of treatment was 49% (19/39) in patients treated with cefiderocol vs. 18% (3/17) in those treated with BAT; however, baseline risk for mortality was higher in the cefiderocol group and BAT was largely heterogenous [22]. The APEKS-NP trial compared cefiderocol with meropenem (high-dose, extended-infusion) for the treatment of Gram-negative nosocomial pneumonia: all-cause mortality in patients with CRAB infection at day 14 were 12.4% (18/145) vs. 11.6% (17/146) in the cefiderocol and meropenem groups, respectively [23].

A prospective, observational, single-centre study compared the clinical failure (defined as the need to switch to second-line antibiotic therapy due to lack of clinical response or recurrence of VAP up to 7 days from the end of active therapy) between cefiderocol and colistin-based regimen groups, in non-COVID-19 ICU patients with CRAB [31]. Cefiderocol was administered as a combination treatment in 52.5% (21/40) of the cases, and as monotherapy in the remaining 19 cases. Multivariable Cox regression analysis showed that a first-line cefiderocol-based regimen was an independent protective factor in clinical failure risk (HR 0.38, 95% CI 0.18–0.76, $p = 0.007$), the result confirmed through the IPTW analysis. However, there was no significant difference in 28-day all-cause mortality between groups: 35% (14/40) vs. 52% (26/50) in cefiderocol and colistin-regimen groups, respectively [31]. Four Italian retrospective studies compared cefiderocol with colistin-based regimen in ICU patients with CRAB [10,11,32,33] (see Table 2)

Table 2. Retrospective observational studies comparing cefiderocol with colistin-based regimen in ICU patients with CRAB.

	Pascale et al. [32] Multicentre (January 2020–April 2021)	**Mazzitelli et al. [33] Single-Centre (August 2020–July 2022)**	**Falcone et al. [10] Single-Centre (January 2020–August 2021)**	**Russo et al. [11] Single-Centre (March 2020–August 2022)**
Population: antibiotic-based regimen groups	107 patients: 42 CFD 65 COL	111 patients: 60 CFD 51 COL	124 patients: 47 CFD 77 COL	73 patients: 19 CFD 54 COL
COVID-19 coinfection	100%	32%	38.7%	100%
Site of infection	BSI (58%) LRTI (41%) Others (1%)	BSI (47.7%) Pneumonia (52.3%)	BSI (57.4%) VAP (25.5%) Others (17%)	VAP and concomitant BSI (100%)
Patients received CFD in combination	0	30 (50%)	33 (70%)	19 (100%)
Main agents co-administered with CFD	/	TGC (18/30) MEM (13/30) FOS (8/30)	TGC (21/33) FOS (8/33)	FOS (7/19) FOS + TGC (7/19) TGC (1/19)
28–30 day all-cause mortality: CFD group vs. COL group	23 (55%) vs. 38 (58%) (*p*-value: 0.7)	26 (51%) vs. 22 (37%) (*p*-value: 0.13)	16 (34%) vs. 43 (56%) (*p*-value: 0.018)	6 (31.5%) vs. 53 (98%) (*p*-value < 0.001)

Legend. ICU: intensive care unit; CRAB: carbapenem-resistant *Acinetobacter baumannii*; CFD: cefiderocol; COL: colistin; BSI: bloodstream infection; LRTI: low respiratory tract infection; VAP: ventilator-associated pneumonia; TGC: tigecycline; MEM: meropenem; FOS: fosfomycin.

Two of the studies demonstrated a comparable rate of all-cause mortality between treatment groups at days 28 and 30, respectively [32,33]. In contrast, the cefiderocol-based regimen was identified as an independent predictor of 30-day survival in the remaining studies [10,11].

In a recent study aimed at evaluating the clinical efficacy of cefiderocol-based regimens for the treatment of CRAB infections, a meta-analysis was conducted on the available data (five abovementioned observational studies and the CREDIBLE-CR trial): a trend was demonstrated towards a significantly lower mortality rate in patients who received a cefiderocol-based compared to a colistin-based regimen [34]. Interestingly, after excluding studies at high/severe risk of bias and considering only studies performing proper adjustments for confounders, cefiderocol-based regimens were associated with a significantly lower risk of mortality (N = 4; OR 0.53; 95% CI 0.39–0.71; I2 = 0.0%) [34]; however, the number of included studies was limited. Unfortunately, no subgroup analysis according to cefiderocol mono- or combination therapy was performed due to a lack of available adjusted data [34]. Anyhow, it is interesting to note that among the mentioned four retrospective studies [10,11,29,30], cefiderocol was an independent predictor of 30-day survival in the studies where the rate of cefiderocol combination-regimen was higher (see Table 2). Furthermore, in the study by Falcone et al., the microbiological failure rate was significantly higher in patients who received monotherapy compared to those who received combination therapy 42.9% (6/14) vs. 6.3% (2/32) and four patients with microbiological failure developed resistance to cefiderocol [10].

The conflicting data between the CREDIBLE-CR trial and observational studies could be explained, and the net of the biases already described, by the fact that in real-world studies, cefiderocol was administered more frequently in a combination treatment. A lower efficacy in monotherapy could in turn be explained by the phenomenon of heteroresistance. Therefore, the real clinical impact of cefiderocol heteroresistance needs to be investigated.

While available data discourage the use of cefiderocol monotherapy in CRAB infections, real-world data on its use as a back-bone agent in combination treatment are encouraging. To accelerate the right placement in therapy of such a compound, priority should be given to studies that address the following questions:

(1) Is there a role for cefiderocol as a back-bone agent in combination treatment for CRAB infections?
(2) What is the best antibiotic partner to use in combination with cefiderocol in CRAB infections?

A trial comparing cefiderocol + ampicillin-sulbactam vs. colistin ± meropenem for CRAB infections is currently registered, although not yet recruiting (NCT05922124) [21].

Of interest, a case of VAP due to extremely-drug-resistant (XDR)-*A. baumannii* successfully treated with cediferocol + sulbactam/durlobactam has been reported [35]. Probably, rather than as a combination partner for sulbactam/durlobactam, cefiderocol might be better used as an alternative back-bone regimen when sulbactam/durlobactam is contraindicated (i.e., resistance or intolerance). As a matter of fact, cefiderocol is the only known option against MBL-CRAB. A cefiderocol-sparing approach could prove to be a winning strategy in case of an increase in the rate of CRAB resistance to sulbactam/durlobactam, an event that is likely to occur after the drug's worldwide commercialization.

Randomized clinical trials (RCTs) to evaluate the efficacy of these two NBLs in combination treatment for CRAB infections are urgently needed.

3. Place in Therapy of Traditional Agents for Treatment of CRAB

3.1. Polymyxins

This antibiotic acts by binding to the anionic molecules of LPS, displacing Mg^{2+} and Ca^{2+} from the outer cell membrane of Gram-negative bacteria, causing permeability changes in the cell envelope and leakage of cell contents. The mechanism of resistance in *A. baumannii* is determined by the complete loss of lipopolysaccharide production.

Colistin, available intravenously as the prodrug colistimethate sodium, and polymyxin B have been the most used therapeutic options for CRAB [36]. Despite colistin resistance being uncommon in *A. baumannii*, there has recently been a worldwide increase in the resistance rate, reaching a peak of approximately 10% in Europe [37].

Polymyxins lung penetration is suboptimal; furthermore, colistin has several disadvantages compared to polymyxin B in terms of pharmacokinetic characteristic: due to the prodrug administration, colistin plasma concentration rises slowly, is subjected to a greater inter-patient variability and, in patients with normal renal function, the target plasma concentration is difficult to achieve [38]. Therefore, except for urinary tract infections (UTIs), polymyxin B is preferred to colistin for severe infections [38]; however, its availability is limited globally. On the contrary, colistin is preferred in UTIs due to its higher urinary concentration (polymyxin B is extensively reabsorbed by the renal tubular cells) [39]; moreover, it seems to present fewer side effect than polymyxin B when administered by inhalation [40]. However, data on the real clinical utility of polymyxins inhalation in patients with CRAB pneumonia are conflicting [40], as are the recommendations of professional societies [8,9,38]. The main adverse effect of polymyxins administration is nephrotoxicity, which in some reports reaches the rate of 55%, while neurotoxicity is less common [40].

Although the combination of colistin with other antibiotics (rifampicin, fosfomycin, meropenem), produced an in vitro synergistic effect against *A. baumannii*, clinical trials failed to demonstrate improved efficacy of colistin in combination with these antibiotics compared to monotherapy [41–43]. Several safer polymyxin candidates, with improved activity (also in term of lung penetration) compared to colistin and polymyxin B, are undergoing preclinical and clinical evaluations [39]. Until their commercialization, it would be appropriate, in our opinion, to use polymyxins as alternative anti-CRAB agents (when no other options are available). In conclusion, the drug in question has been found to have certain limitations in terms of its use. These include nephrotoxicity [40], suboptimal pulmonary penetration, and suboptimal plasma concentrations [38]. It is regarded as a potential alternative agent in instances where no other viable options exist.

3.2. Tetracycline Derivatives

The pharmacological action of these drugs is exerted through the inhibition of the 30S ribosomal subunit, thereby impeding protein synthesis. *A. baumannii* resistance is mediated by three distinct mechanisms: (i) ATP-dependent efflux, (ii) inactivation of tetracyclines by enzymes, and (iii) ribosomal protection proteins (RPPs).

Tigecycline and minocycline, capable of escaping common tetracycline resistance mechanisms, are currently recommended for CRAB infections treatment [7,8]. Clinical breakpoints for tigecycline against *Acinetobacter* spp. have not been established [8], while the susceptibility rate to minocycline is about 85%, dropping to around 70% for multi-drug resistant (MDR) isolates [44].

Although minocycline seems to have better lung penetration than tigecycline [45,46], both compounds are characterized by suboptimal exposures in blood and serum [5]. Therefore, high-dose administration is recommended for CRAB infections [7,8]. However, in a PTA analysis, the high-dose minocycline regimen currently employed in clinical practice was predicted to result in a suboptimal plasma fAUC:MIC profile for patients with *A. baumannii* infections with MICs > 1 mg/L; among MDR-ABC, minocycline MIC values were >1 mg/L for 60% of the tested isolates [45]. Such data would call into question the current susceptibility breakpoint for minocycline ($\leq$4 mg/L), based on the rat pneumonia model [45]. Furthermore, the global availability of intravenous minocycline formulation is limited. Regarding the PK/PD data on high-dose tigecycline, a study on serum and ELF concentrations among critically ill patients was conducted: PK/PD target attainment for pneumonia was $\geq$75% with MICs $\leq$ 0.5 mcg/mL [46], but only 31% of international CRAB isolates demonstrated tigecycline MICs $\leq$ 0.5 [33].

However, in vitro synergism between tigecycline and cefiderocol has been demonstrated among both cefiderocol-resistant and susceptible CRAB isolates [47]; moreover, in observational studies, tigecycline was one of the most frequently administered agents in a cefiderocol-combination regimen [34]. Tigecycline (or minocycline where iv formulation is available) could represent a suitable partner for both NBLs in CRAB infections treatment. RCTs are needed to confirm this hypothesis.

Preclinical and clinical data on the novel tetracycline derivatives, eravacycline and omadacycline, suggest reduced activity against CRAB compared to tigecycline and minocycline [8]; however, data are scarce and further studies are necessary to understand if and what role they may have in CRAB infections. It is important to note that suboptimal efficacy has been observed in serum, lung, and urine samples [5].

3.3. Fosfomycin

Fosfomycin exerts its antimicrobial effect by inhibiting the synthesis of peptidoglycan, a component of the bacterial cell wall, at an earlier stage than betalactams. *A. baumannii* develops resistance to fosfomycin through different mechanisms, including the presence of the fosfomycin efflux transporter MFS-encoded AbaF and fosfomycin resistance glutathione transferase.

Fosfomycin is characterized by good tissue penetration (including infected lung tissue) as well as a good concentration in serum, urine, and cerebrospinal fluid, with a good safety profile [48]. ABC is intrinsically resistant to fosfomycin; however, it resulted in a synergistic effect in vitro when combined with several antibiotics against CRAB [48,49].

Due to a single trial which failed to demonstrate the superiority of colistin plus fosfomycin over colistin alone [43], but also because intravenous fosfomycin is not available in the U.S., its role in CRAB treatment has been under-investigated [49] and the drug is not currently recommended for such infections [7,8]. Regardless, in a prospective, observational, multicentre study, conducted on 180 patients with HAP due to MDR—*A. baumannii*, 44 patients were treated with a fosfomycin-containing regimen (29 in double and 15 in triple combination regimen) which turned out to be a factor associated with 30-day survival ($p < 0.001$) [50]. Moreover, in a case-series study conducted on 20 ICU patients with BSI due to pan-drug resistant (PDR)—*A. baumannii*, a fosfomycin-containing regimen (one case in double regimen and seven cases in regimens including at least three antibiotics) was associated with 28-day survival ($p < 0.005$) [51].

Regarding the potential role as a partner for NBLs, in vitro data showed a synergistic effect of fosfomycin in combination with cefiderocol against a cefiderocol-resistant CRAB isolate [52]. Furthermore, in observational studies, fosfomycin was the most frequently administered agent in a cefiderocol-combination regimen [34].

Interestingly, in one of the abovementioned retrospective studies, fosfomycin was co-administered in 14 out 19 patients of the cefiderocol group, resulting an independent factor of 30-day survival ($p < 0.001$); however, 8 out of 14 patients received fosfomycin in a combination of at least three antibiotics [11].

Finally, a study conducted on sulbactam-resistant CRAB isolates in a hollow-fiber infection model showed that a combination of fosfomycin and extended infusion of sulbactam produced a 4 $\log_{10}$ reduction in colony count within 24 h, followed by suppression of regrowth [53]. Although fosfomycin is not currently recommended, it could represent a suitable partner for both NBLs in CRAB infections treatment. RCTs are needed to confirm this hypothesis.

3.4. High-Dose Extended-Infusion Meropenem

A high-dose extended-infusion meropenem regimen enhances the PK exposure of the compound, but considering the high meropenem MIC values of CRAB isolates, it does not reach the optimal cumulative fraction of response in such infections [54]. However, a triple combination regimen containing meropenem has been used successfully for XDR-CRAB infections [5] and, in one case, meropenem was co-administered with cefiderocol [55].

Interestingly, cefiderocol + meropenem demonstrated an in vitro synergistic effect against 79.5% of cefiderocol-resistant isolates [47]. A case of PDR-CRAB infection was cured with a combination of sulbactam/durlobactam + meropenem; of note, the addition of meropenem reduced the sulbactam/durlobactam MIC from 8 to 4 mg/L (the preliminary susceptibility breakpoint) [56]. If further studies confirm the synergism between meropenem and NBLs, high-dose extended-infusion meropenem + sulbactam/durlobactam (or cefiderocol) could represent a valuable option for PDR-CRAB infections.

From this perspective, a meropenem-sparing strategy should be adopted in case of CRAB infections due to strains sensitive to the NBLs. In case of co-administration of high-dose meropenem with another β-lactam, a close monitoring of side effects should be warranted, due to the high probability of an increase in side effects (i.e., epilepsy).

3.5. Aminoglycosides

These antibiotics act by binding to the 30S subunit of ribosomes, thereby inhibiting protein synthesis in bacteria. The resistance mechanism results in enzymatic modification of the aminoglycoside molecule, primarily through *N*-acetylation, *O*-nucleotidylation, or *O*-phosphorylation at various points along the molecule.

Aminoglycosides are currently recommended in combination treatment for susceptible CRAB isolates [7]; however, the global resistance rate is >80% [37]. Furthermore, considering drugs' side effects [57] and suboptimal ELF concentrations [58], their prospective role in CRAB infections will likely be limited to a few selected cases.

3.6. *Rifamycins*

Rifamycins act as inhibitors of bacterial DNA-dependent RNA polymerase. *A. baumannii* resistance is the result of mutations in the rpoB gene, which encodes the beta-subunit of rifamycin-sensitive RNA polymerase. These mutations prevent RNA elongation immediately after the addition of the first nucleotides.

Rifampicin showed synergistic effect against MDR-*A. baumannii* when combined with colistin [59], but the clinical efficacy of such treatment has not been demonstrated [42] and rifampicin is currently not recommended in CRAB infections [7,8]. However, in vitro synergism seems to depend by rifampicin MICs, and data on rifampicin MIC values of CRAB isolates are scant [59].

Interestingly, in vitro synergism between rifampicin and sulbactam against CRAB isolates has been demonstrated [60]; moreover, in a case series on 12 infant and young children with severe VAP caused by XDR-*A. baumannii*, the combination of rifampicin and sulbactam appeared to be an effective and safe therapy (9 out of 12 patients were considered cured) [61]. Studies addressing an in vitro synergic effect between rifampicin and the NBLs should be conducted, eventually followed by clinical evaluations of combined treatments.

Rifabutin, available only as an oral formulation, overcomes the common rifamycin resistance mechanisms of *A. baumannii*, displaying potent in vitro activity against CRAB strains [62]. The drug is rapidly distributed in all organs and tissues, where levels are constantly higher than plasma levels; unfortunately, the oral bioavailability of rifabutin is very low, limiting its clinical utility [62]. An intravenous formulation of rifabutin (BV100) has been developed to maximize clinical efficacy against *A. baumannii* infections while minimizing the risk of resistance development, and it is currently under investigation in a Phase I clinical trial [62].

3.7. *Trimethoprim/Sulfamethoxazole*

Trimethoprim is a dihydrofolate reductase (DHFR) inhibitor, which blocks the formation of tetrahydrofolic acid by dihydrofolic acid. Sulfonamides, on the other hand, are known to inhibit dihydropteroate synthase (DHPS). The presence of trimethoprim-resistant dihydrofolate reductase A. baumannii is considered resistant.

Sporadic cases of MDR—*A. baumannii* have been successfully treated with a combination regimen including trimethoprim/sulfamethoxazole [11,63], while in a retrospective match cohort study, comparing trimethoprim/sulfamethoxazole monotherapy with other regimens for CRAB infections, all-cause 30-day mortality was lower in trimethoprim/sulfamethoxazole group (24.5%, 13 of 53 vs. 38.6%, 32 of 83); however, the baseline risk for mortality was higher in the comparator antibiotics group [64]. The resistance rate among CRAB isolates is estimated to be >80% [63] and trimethoprim/sulfamethoxazole is currently not recommended for such infections [8,9].

Possible synergistic effects between trimethoprim/sulfamethoxazole and the NBLs should be investigated; the compound could represent a valuable partner in CRAB combination-treatment, at least for infections due to sensitive isolates.

3.8. *Novel Antibiotics (Zosurazalpin)*

Among the most interesting and promising options not yet on the market, zosurazalpin has been identified as a drug with potential antibacterial activity against CRAB in vitro and in mouse models, capable of overcoming resistance mechanisms. The mechanism of action involves inhibition of the LptB2FGC complex by blocking the transport of bacterial lipopolysaccharide from the inner membrane to its target on the outer membrane.

The main characteristics of antibiotics are summarized in Table 3.

Table 3. Potential role of anti-CRAB antibiotics.

	Potential Role	Main Mechanisms of Action	Main Mechanisms of Resistance	Evidences (or Available Data)	Limits	Studies to Be Prioritized
Sulbactam/durlobactam	Back-bone agent in combination treatment	Inhibition of penicillin binding proteins 1 and 3 (involved in synthesis of bacterial peptidoglycan)/Uses a reversible mechanism of inhibition through β-lactamase active site carbamoylation	Single amino acid changes near the active site serine of PBP3 (S336), the target of sulbactam	RCT: non-inferior to COL (both co-administered with IPM-CLN) [17]	Efficacy as monotherapy not known	RCTs finding the best partner-agent
Cefiderocol	Back-bone agent in combination treatment	Utilizes the siderophore–iron complex pathway to penetrate the outer membrane of Gram-negative organisms in addition to normal passive diffusion through membrane porins	Alterations of the intrinsic AmpC and siderophore receptors	Metanalysis: lower risk of mortality rate compared to COL-based regimen [34]	Unsatisfactory efficacy as monotherapy when compared to COL [22] and MEM [23]	-RCTs confirming the role as back-bone agent; -RCTs finding the best partner-agent
Polymyxins	COL (or PB): alternative agent (when no other options are available)	Binds with the anionic LPS molecules by displacing Mg^{2+} and Ca^{2+} from the outer cell membrane of Gram-negative bacteria, leading to permeability changes in the cell envelope and leakage of cell contents	Complete loss of Lipopolysaccharide production	Large clinical experience as back-bone agent [36]. (Data on combination with NBLs are missing.)	-Nefrotoxicity [40]; -suboptimal lung penetration [38]; -suboptimal plasma concentrations [38].	Accelerate studies on safer polymyxin with lung improved activity
Tetracycline derivatives	High-dose TGC (or MNC): partner-agent in combination treatment	Inhibit the 30S ribosomal subunit and thereby inhibit protein synthesis	(i) Efflux dependent on ATP, (ii) inactivation of tetracyclines by enzymes, and (iii) ribosomal protection proteins (RPPs)	TGC + CFD: -in vitro synergism [47]. (one of the most frequently used combination in observational studies [34])	Suboptimal exposures in serum, lung and urine [5]	RCTs comparing TGC and FOS as partner-agent
Fosfomycin	Partner-agent in combination treatment	Inhibition of bacterial cell wall peptidoglycan synthesis at an earlier stage than betalactams.	(i) fosfomycin efflux MFS transporter AbaF encoded; (ii) fosfomycin resistance glutathione transferases	-Retrospective study: associated with 30-day survival in combination with CFD [11]; -In vitro synergism with CFD [52] and SUL [53]; (the most commonly used agent in combination with CFD in observational studies [34])	-Data coming from the observational study included regimens of more than 2 agents [11] -*A. baumannii* is intrinsically resistant to the drug [48]	RCTs comparing TGC and FOS as partner-agent

Table 3. *Cont.*

	Potential Role	Main Mechanisms of Action	Main Mechanisms of Resistance	Evidences (or Available Data)	Limits	Studies to Be Prioritized
High-dose extended-infusion meropenem	Partner-agent in PDR-CRAB infections (to be spared in treatment of strains sensitives to NBLs)	Binds penicillin-binding protein (PBP) in the bacterial cell wall and inhibits peptidoglycan cross-linking associated with cell wall synthesis	Production of enzymes such as beta-lactamases	In vitro synergism against CFD-resistant strains [47]. (Combined with SUL-DUR: a single case report of PDR-CRAB cured, with in vitro synergistic effect [56])	-Suboptimal cumulative fraction of response [54]; -possible increase in side effects rate if co-administered with other BLs	-In vitro studies on synergism with NBLs; -Clinical studies on PDR-CRAB infections
Aminoglycosides	Alternative partner-agent for few selected cases	Bind to the 30S subunit of ribosomes, inhibiting protein synthesis in bacteria.	Enzymatic modification primarily through N-acetylation, O-nucleotidylation, or O-phosphorylation at different locations of the aminoglycoside molecule.	Currently recommended as a combination treatment for susceptible CRAB isolates [7]. (Data on combination with NBLs are missing)	-Resistance rate among CRAB isolates > 80%; -suboptimal concentration in lung [37]; -high rate side effects [57]	/
Rifamycins	Alternative partner-agent	Inhibit bacterial DNA-dependent RNA polymerase	Mutations in the rpoB gene, which encodes rifamycin sensitive beta-subunit of RNA polymerase and averts RNA elongation just after adding the first nucleotides	RFM + SUL: -in vitro synergism [60]; (a case series on 12 pediatric patients reported clinical efficacy in VAP due to XDR-*A. baumannii* [61])	Synergism seems to depend by rifampicin MICs, but MICs data are scant [59]	-Accelerate clinical studies on rifabutin iv formulation; -in vitro studies on synergism between RFM and NBLs
Trimethoprim/sulfamethoxa-zole	Alternative partner-agent	Trimethoprim is a dihydrofolate reductase (DHFR) inhibitor (blocking tetrahydrofolic acid formation by dihydrofolic acid), while sulfonamides are known dihydropteroate synthase (DHPS) inhibitors	Trimethoprim-resistant dihydrofolate reductases	(Successfully administered in combination with CFD in sporadic cases [11])	Resistance rate among CRAB isolates > 80% [63]	In vitro studies on synergism with NBLs

Legend. CRAB: carbapenem-resistant *Acinetobacter baumannii*; RCT: randomized clinical trial; COL: colistin; IMP-CLN: imipenem/cilastatin; MEM: meropenem; PB: polymyxin B; NBLs: newly developed β-lactam agents; TGC: tigecycline; MNC: minocycline; CFD: cefiderocol; FOS: fosfomycin; SUL: sulbactam; *A. baumannii*: *Acinetobacter baumannii*; PDR: pan-drug resistant; DUR: durlobactam; BLs: β-lactam agents; RFM: rifampicin; VAP: ventilator-associated pneumonia; XDR: extremely drug resistant; MICs: minimal inhibitory concentrations. High-dose tigecycline: 200 mg as a loading dose followed by 100 mg q12h; high-dose minocycline: 200 mg q12h; high-dose extended-infusion meropenem: 2 g over 3 hrs. infusion q8h.

4. Conclusions

The commercialization of NBLs opens a new scenario for the treatment of CRAB infections.

However, data are limited, and relevant clinical questions remain outstanding, such as the superiority of a combination treatment over monotherapy and the best antibiotic partner to use in combination.

In this narrative review, we have highlighted the main characteristics and potential roles of anti-CRAB antibiotics (summarized in Table 3), in view of recent data on NBLs. Sulbactam/durlobactam seems to be the best candidate to replace current back-bone agents. Cefiderocol, despite the suboptimal efficacy in monotherapy, could play a crucial role in combination-regimen (probably as an alternative back-bone agent): RCTs are needed to better evaluate the role of these new β-lactams in CRAB infections. Due to toxicity and the PK/PD limitations, colistin should be used as an alternative anti-CRAB agent (when no other options are available).

Tigecycline and fosfomycin could represent suitable partners for both NBLs: RCTs –comparing the efficacy of these two drugs as partner-antibiotics in combined CRAB treatment should be prioritized [65]. High-dose extended-infusion meropenem, if further studies will confirm the synergistic effect between meropenem and NBLs' should be taken into consideration in case of PDR-CRAB infections, while the role of aminoglycosides will be probably limited to a few selected cases. In vitro data on the synergism between NBLs and rifampicin or trimethoprim/sulfamethoxazole are needed to guide eventual any future clinical investigations into the possible role of these "old antibiotics" as partner-agents in CRAB infection treatment.

Taking in account the limitations of current available partner-antibiotic candidates for CRAB infections, huge efforts should be made to accelerate pre-clinical and clinical studies on safer polymyxin candidates with improved lung activity, as well as on the intravenous formulation of rifabutin or about new promising molecules like cefoperazone-sulbactam [66]. Important strategies about carbapenem-sparing could be another important point to reduce the spread of CRAB strains [67].

Author Contributions: F.S. and M.G. conceptualized and wrote the manuscript; S.P.G. and R.L. wrote the manuscript; A.B., E.G., F.L., A.R. and E.M.T. revised the manuscript. All authors have read and agreed to the published version of the manuscript.

Funding: This research received no external funding.

Conflicts of Interest: The authors declare no conflicts of interest.

References

1. Peleg, A.Y.; Seifert, H.; Paterson, D.L. *Acinetobacter baumannii*: Emergence of a successful pathogen. *Clin. Microbiol. Rev.* **2008**, *21*, 538–582. [CrossRef] [PubMed]
2. Wong, D.; Nielsen, T.B.; Bonomo, R.A.; Pantapalangkoor, P.; Luna, B.; Spellberg, B. Clinical and pathophysiological overview of *Acinetobacter* infections: A century of challenges. *Clin. Microbiol. Rev.* **2017**, *30*, 409–447. [CrossRef] [PubMed]
3. Harding, C.M.; Hennon, S.W.; Feldman, M.F. Uncovering the mechanisms of *Acinetobacter baumannii* virulence. *Nat. Rev. Microbiol.* **2018**, *16*, 91–102. [CrossRef] [PubMed]
4. Castanheira, M.; Mendes, R.E.; Gales, A.C. Global Epidemiology and Mechanisms of Resistance of *Acinetobacter baumannii-calcoaceticus* Complex. *Clin. Infect. Dis.* **2023**, *76* (Suppl. S2), S166–S178. [CrossRef] [PubMed]
5. Shields, R.K.; Paterson, D.L.; Tamma, P.D. Navigating Available Treatment Options for Carbapenem-Resistant *Acinetobacter baumannii-calcoaceticus* Complex Infections. *Clin. Infect. Dis.* **2023**, *76* (Suppl. S2), S179–S193. [CrossRef] [PubMed]
6. Antimicrobial Resistance Collaborators. Global burden of bacterial antimicrobial resistance in 2019: A systematic analysis. *Lancet* **2022**, *399*, 629–655. [CrossRef] [PubMed]
7. Paul, M.; Carrara, E.; Retamar, P.; Tängdén, T.; Bitterman, R.; Bonomo, R.A.; de Waele, J.; Daikos, G.L.; Akova, M.; Harbarth, S.; et al. European Society of Clinical Microbiology and Infectious Diseases (ESCMID) guidelines for the treatment of infections caused by multidrug-resistant Gram-negative bacilli (endorsed by European society of intensive care medicine). *Clin. Microbiol. Infect.* **2022**, *28*, 521–547. [CrossRef] [PubMed]
8. Tamma, P.D.; Aitken, S.L.; Bonomo, R.A.; Mathers, A.J.; van Duin, D.; Clancy, C.J. Infectious Diseases Society of America 2023 Guidance on the Treatment of Antimicrobial Resistant Gram-Negative Infections. *Clin. Infect. Dis.* **2023**, *18*, ciad428. [CrossRef]

9. US Food and Drug Administration. FDA Approves New Treatment for Pneumonia Caused by Certain Difficult-to-Treat Bacteria [Media Release]. 23 May 2023. Available online: https://www.fda.gov/news-events/press-announcements/fda-approves-new-treatment-pneumonia-caused-certain-difficult-treat-bacteria (accessed on 2 May 2024).
10. Falcone, M.; Tiseo, G.; Leonildi, A.; Della Sala, L.; Vecchione, A.; Barnini, S.; Farcomeni, A.; Menichetti, F. Cefiderocol- compared to colistin-based regimens for the treatment of severe infections caused by carbapenem-resistant *Acinetobacter baumannii*. *Antimicrob. Agents Chemother.* **2022**, *66*, e0214221. [CrossRef]
11. Russo, A.; Bruni, A.; Gullì, S.; Borrazzo, C.; Quirino, A.; Lionello, R.; Serapide, F.; Garofalo, E.; Serraino, R.; Romeo, F.; et al. Efficacy of cefiderocol- vs colistin-containing regimen for treatment of bacteraemic ventilator-associated pneumonia caused by carbapenem-resistant *Acinetobacter baumannii* in patients with COVID-19. *Int. J. Antimicrob. Agents.* **2023**, *62*, 106825. [CrossRef]
12. Penwell, W.F.; Shapiro, A.B.; Giacobbe, R.A.; Gu, R.-F.; Gao, N.; Thresher, J.; McLaughlin, R.E.; Huband, M.D.; DeJonge, B.L.M.; Ehmann, D.E.; et al. Molecular mechanisms of sulbactam antibacterial activity and resistance determinants in *Acinetobacter baumannii*. *Antimicrob. Agents Chemother.* **2015**, *59*, 1680–1689. [CrossRef] [PubMed]
13. Kuo, S.-C.; Lee, Y.-T.; Lauderdale, T.-L.Y.; Huang, W.-C.; Chuang, M.-F.; Chen, C.-P.; Su, S.-C.; Lee, K.-R.; Chen, T.-L. Contribution of Acinetobacter-derived cephalosporinase-30 to sulbactam resistance in *Acinetobacter baumannii*. *Front. Microbiol.* **2015**, *6*, 231. [CrossRef]
14. Durand-Réville, T.F.; Guler, S.; Comita-Prevoir, J.; Chen, B.; Bifulco, N.; Huynh, H.; Lahiri, S.; Shapiro, A.B.; McLeod, S.M.; Carter, N.M.; et al. ETX2514 is a broad-spectrum β-lactamase inhibitor for the treatment of drug-resistant Gram-negative bacteria including *Acinetobacter baumannii*. *Nat. Microbiol.* **2017**, *2*, 17104. [CrossRef] [PubMed]
15. Principe, L.; Di Bella, S.; Conti, J.; Perilli, M.; Piccirilli, A.; Mussini, C.; Decorti, G. *Acinetobacter baumannii* Resistance to Sulbactam/Durlobactam: A Systematic Review. *Antibiotics* **2022**, *11*, 1793. [CrossRef] [PubMed]
16. Rodvold, K.A.; Gotfried, M.H.; Isaacs, R.D.; O'Donnell, J.P.; Stone, E. Plasma and Intrapulmonary Concentrations of ETX2514 and Sulbactam following Intravenous Administration of ETX2514SUL to Healthy Adult Subjects. *Antimicrob. Agents Chemother.* **2018**, *62*, e01089-18. [CrossRef]
17. Kaye, K.S.; Shorr, A.F.; Wunderink, R.G.; Du, B.; Poirier, G.E.; Rana, K.; Miller, A.; Lewis, D.; O'Donnell, J.; Chen, L.; et al. Efficacy and safety of sulbactam-durlobactam versus colistin for the treatment of patients with serious infections caused by *Acinetobacter baumannii-calcoaceticus* complex: A multicentre, randomised, active-controlled, phase 3, non-inferiority clinical trial (ATTACK). *Lancet Infect. Dis.* **2023**, *23*, 1072–1084. [PubMed]
18. Giuliano, S.; Sbrana, F.; Tascini, C. Sulbactam-durlobactam for infections caused by *Acinetobacter baumannii-calcoaceticus* complex. *Lancet Infect. Dis.* **2023**, *23*, e274. [CrossRef] [PubMed]
19. Kaye, K.S.; McLeod, S.M.; O'Donnell, J.P.; Altarac, D. Sulbactam-durlobactam for infections caused by *Acinetobacter baumannii-calcoaceticus* complex—Authors' reply. *Lancet Infect. Dis.* **2023**, *23*, e275–e276. [CrossRef] [PubMed]
20. Petropoulou, D.; Siopi, M.; Vourli, S.; Pournaras, S. Activity of Sulbactam-Durlobactam and Comparators Against a National Collection of Carbapenem-Resistant *Acinetobacter baumannii* Isolates From Greece. *Front. Cell Infect. Microbiol.* **2022**, *11*, 814530. [CrossRef]
21. National Library of Medicine. ClinicalTrials.gov. Available online: https://clinicaltrials.gov/ (accessed on 2 May 2024).
22. Bassetti, M.; Echols, R.; Matsunaga, Y.; Ariyasu, M.; Doi, Y.; Ferrer, R.; Lodise, T.P.; Naas, T.; Niki, Y.; Paterson, D.L.; et al. Efficacy and safety of cefiderocol or best available therapy for the treatment of serious infections caused by carbapenem-resistant Gram-negative bacteria (CREDIBLE-CR): A randomised, open-label, multicentre, pathogen-focused, descriptive, phase 3 trial. *Lancet Infect. Dis.* **2021**, *21*, 226–240. [CrossRef]
23. Wunderink, R.G.; Matsunaga, Y.; Ariyasu, M.; Clevenbergh, P.; Echols, R.; Kaye, K.S.; Kollef, M.; Menon, A.; Pogue, J.M.; Shorr, A.F.; et al. Cefiderocol versus high-dose, extended-infusion meropenem for the treatment of Gram-negative nosocomial pneumonia (APEKS-NP): A randomised, double-blind, phase 3, non-inferiority trial. *Lancet Infect. Dis.* **2021**, *21*, 213–225. [CrossRef] [PubMed]
24. Syed, Y.Y. Cefiderocol: A Review in Serious Gram-Negative Bacterial Infections. *Drugs.* **2021**, *81*, 1559–1571. [CrossRef] [PubMed]
25. Kollef, M.; Dupont, H.; Greenberg, D.E.; Viale, P.; Echols, R.; Yamano, Y.; Nicolau, D.P. Prospective role of cefiderocol in the management of carbapenem-resistant *Acinetobacter baumannii* infections: Review of the evidence. *Int. J. Antimicrob. Agents.* **2023**, *62*, 106882. [CrossRef] [PubMed]
26. Karlowsky, J.A.; Hackel, M.A.; Takemura, M.; Yamano, Y.; Echols, R.; Sahm, D.F. In Vitro Susceptibility of Gram-Negative Pathogens to Cefiderocol in Five Consecutive Annual Multinational SIDERO-WT Surveillance Studies, 2014 to 2019. *Antimicrob. Agents Chemother.* **2022**, *66*, e0199021. [CrossRef] [PubMed]
27. Poirel, L.; Sadek, M.; Nordmann, P. Contribution of PER-Type and NDM-Type β-Lactamases to Cefiderocol Resistance in *Acinetobacter baumannii*. *Antimicrob. Agents Chemother.* **2021**, *65*, e0087721. [CrossRef]
28. Choby, J.E.; Ozturk, T.; Satola, S.W.; Jacob, J.T.; Weiss, D.S. Widespread cefiderocol heteroresistance in carbapenem-resistant Gram-negative pathogens. *Lancet Infect. Dis.* **2021**, *21*, 597–598. [CrossRef] [PubMed]
29. Karakonstantis, S.; Rousaki, M.; Kritsotakis, E.I. Cefiderocol: Systematic Review of Mechanisms of Resistance, Heteroresistance and In Vivo Emergence of Resistance. *Antibiotics* **2022**, *11*, 723. [CrossRef] [PubMed]
30. Kawaguchi, N.; Katsube, T.; Echols, R.; Wajima, T. Population Pharmacokinetic and Pharmacokinetic/Pharmacodynamic Analyses of Cefiderocol, a Parenteral Siderophore Cephalosporin, in Patients with Pneumonia, Bloodstream Infection/Sepsis, or Complicated Urinary Tract Infection. *Antimicrob. Agents Chemother.* **2021**, *65*, e01437-20. [CrossRef]

31. Dalfino, L.; Stufano, M.; Bavaro, D.F.; Diella, L.; Belati, A.; Stolfa, S.; Romanelli, F.; Ronga, L.; Di Mussi, R.; Murgolo, F.; et al. Effectiveness of First-Line Therapy with Old and Novel Antibiotics in Ventilator-Associated Pneumonia Caused by Carbapenem-Resistant *Acinetobacter baumannii*: A Real Life, Prospective, Observational, Single-Center Study. *Antibiotics* **2023**, *12*, 1048. [CrossRef]
32. Pascale, R.; Pasquini, Z.; Bartoletti, M.; Caiazzo, L.; Fornaro, G.; Bussini, L.; Volpato, F.; Marchionni, E.; Rinaldi, M.; Trapani, F.; et al. Cefiderocol treatment for carbapenem-resistant *Acinetobacter baumannii* infection in the ICU during the COVID-19 pandemic: A multicentre cohort study. *JAC Antimicrob. Resist.* **2021**, *3*, dlab174. [CrossRef]
33. Mazzitelli, M.; Gregori, D.; Sasset, L.; Trevenzoli, M.; Scaglione, V.; Lo Menzo, S.; Marinello, S.; Mengato, D.; Venturini, F.; Tiberio, I.; et al. Cefiderocol-Based versus Colistin-Based Regimens for Severe Carbapenem-Resistant *Acinetobacter baumannii* Infections: A Propensity Score-Weighted, Retrospective Cohort Study during the First Two Years of the COVID-19 Pandemic. *Microorganisms* **2023**, *11*, 984. [CrossRef] [PubMed]
34. Gatti, M.; Cosentino, F.; Giannella, M.; Viale, P.; Pea, F. Clinical efficacy of cefiderocol-based regimens in patients affected by carbapenem-resistant *Acinetobacter baumannii* infections: A systematic review with meta-analysis. *Int. J. Antimicrob. Agents.* **2024**, *63*, 107047. [CrossRef] [PubMed]
35. Zaidan, N.; Hornak, J.P.; Reynoso, D. Extensively Drug-Resistant *Acinetobacter baumannii* Nosocomial Pneumonia Successfully Treated with a Novel Antibiotic Combination. *Antimicrob. Agents Chemother.* **2021**, *65*, e0092421. [CrossRef] [PubMed]
36. Lyu, C.; Zhang, Y.; Liu, X.; Wu, J.; Zhang, J. Clinical efficacy and safety of polymyxins based versus non-polymyxins based therapies in the infections caused by carbapenem-resistant *Acinetobacter baumannii*: A systematic review and meta-analysis. *BMC Infect. Dis.* **2020**, *20*, 296. [CrossRef] [PubMed]
37. Seifert, H.; Blondeau, J.; Lucaßen, K.; Utt, E.A. Global update on the in vitro activity of tigecycline and comparators against isolates of *Acinetobacter baumannii* and rates of resistant phenotypes (2016–2018). *J. Glob. Antimicrob. Resist.* **2022**, *31*, 82–89. [CrossRef]
38. Tsuji, B.T.; Pogue, J.M.; Zavascki, A.P.; Paul, M.; Daikos, G.L.; Forrest, A.; Giacobbe, D.R.; Viscoli, C.; Giamarellou, H.; Karaiskos, I.; et al. International Consensus Guidelines for the Optimal Use of the Polymyxins: Endorsed by the American College of Clinical Pharmacy (ACCP), European Society of Clinical Microbiology and Infectious Diseases (ESCMID), Infectious Diseases Society of America (IDSA), International Society for Anti-infective Pharmacology (ISAP), Society of Critical Care Medicine (SCCM), and Society of Infectious Diseases Pharmacists (SIDP). *Pharmacotherapy* **2019**, *39*, 10–39. [PubMed]
39. Nang, S.C.; Azad, M.A.K.; Velkov, T.; Zhou, Q.T.; Li, J. Rescuing the Last-Line Polymyxins: Achievements and Challenges. *Pharmacol. Rev.* **2021**, *73*, 679–728. [CrossRef]
40. Kassamali, Z.; Jain, R.; Danziger, L.H. An update on the arsenal for multidrug-resistant *Acinetobacter* infections: Polymyxin antibiotics. *Int. J. Infect. Dis.* **2015**, *30*, 125–132. [CrossRef] [PubMed]
41. Paul, M.; Daikos, G.L.; Durante-Mangoni, E.; Yahav, D.; Carmeli, Y.; Benattar, Y.D.; Skiada, A.; Andini, R.; Eliakim-Raz, N.; Nutman, A.; et al. Colistin alone versus colistin plus meropenem for treatment of severe infections caused by carbapenem-resistant Gram-negative bacteria: An open-label, randomised controlled trial. *Lancet Infect. Dis.* **2018**, *18*, 391–400. [CrossRef]
42. Durante-Mangoni, E.; Signoriello, G.; Andini, R.; Mattei, A.; De Cristoforo, M.; Murino, P.; Bassetti, M.; Malacarne, P.; Petrosillo, N.; Galdieri, N.; et al. Colistin and rifampicin compared with colistin alone for the treatment of serious infections due to extensively drug-resistant *Acinetobacter baumannii*: A multicenter, randomized clinical trial. *Clin. Infect. Dis.* **2013**, *3*, 349–358. [CrossRef]
43. Sirijatuphat, R.; Thamlikitkul, V. Preliminary study of colistin versus colistin plus fosfomycin for treatment of carbapenem-resistant *Acinetobacter baumannii* infections. *Antimicrob. Agents Chemother.* **2014**, *9*, 5598–5601. [CrossRef]
44. Flamm, R.K.; Shortridge, D.; Castanheira, M.; Sader, H.S.; Pfaller, M.A. In Vitro Activity of Minocycline against U.S. Isolates of *Acinetobacter baumannii-Acinetobacter calcoaceticus* Species Complex, *Stenotrophomonas maltophilia*, and *Burkholderia cepacia* Complex: Results from the SENTRY Antimicrobial Surveillance Program, 2014 to 2018. *Antimicrob. Agents Chemother.* **2019**, *63*, e01154-19. [PubMed]
45. Lodise, T.P.; Van Wart, S.; Sund, Z.M.; Bressler, A.M.; Khan, A.; Makley, A.T.; Hamad, Y.; Salata, R.A.; Silveira, F.P.; Sims, M.D.; et al. Pharmacokinetic and Pharmacodynamic Profiling of Minocycline for Injection following a Single Infusion in Critically Ill Adults in a Phase IV Open-Label Multicenter Study (ACUMIN). *Antimicrob. Agents Chemother.* **2021**, *65*, e01809-20. [CrossRef] [PubMed]
46. De Pascale, G.; Lisi, L.; Ciotti, G.M.P.; Vallecoccia, M.S.; Cutuli, S.L.; Cascarano, L.; Gelormini, C.; Bello, G.; Montini, L.; Carelli, S.; et al. Pharmacokinetics of high-dose tigecycline in critically ill patients with severe infections. *Ann. Intensive Care.* **2020**, *10*, 94. [CrossRef] [PubMed]
47. Ni, W.; Wang, Y.; Ma, X.; He, Y.; Zhao, J.; Guan, J.; Li, Y.; Gao, Z. In vitro and in vivo efficacy of cefiderocol plus tigecycline, colistin, or meropenem against carbapenem-resistant *Acinetobacter baumannii*. *Eur. J. Clin. Microbiol. Infect. Dis.* **2022**, *41*, 1451–1457. [CrossRef] [PubMed]
48. Falagas, M.E.; Vouloumanou, E.K.; Samonis, G.; Vardakas, K.Z. Fosfomycin. *Clin. Microbiol. Rev.* **2016**, *29*, 321–347. [CrossRef] [PubMed]
49. Guastalognamo, M.; Trecarichi, E.M.; Russo, A. Intravenous fosfomycin: The underdog player in the treatment of carbapenem-resistant *Acinetobacter baumannii* infections. *Clin. Infect. Dis.* **2023**, *77*, 1736–1737. [CrossRef]

50. Russo, A.; Bassetti, M.; Bellelli, V.; Bianchi, L.; Marincola Cattaneo, F.; Mazzocchetti, S.; Paciacconi, E.; Cottini, F.; Schiattarella, A.; Tufaro, G.; et al. Efficacy of a Fosfomycin-Containing Regimen for Treatment of Severe Pneumonia Caused by Multidrug-Resistant *Acinetobacter baumannii*: A Prospective, Observational Study. *Infect. Dis. Ther.* **2021**, *10*, 187–200. [CrossRef]
51. Assimakopoulos, S.F.; Karamouzos, V.; Eleftheriotis, G.; Lagadinou, M.; Bartzavali, C.; Kolonitsiou, F.; Paliogianni, F.; Fligou, F.; Marangos, M. Efficacy of Fosfomycin-Containing Regimens for Treatment of Bacteremia Due to Pan-Drug Resistant *Acinetobacter baumannii* in Critically Ill Patients: A Case Series Study. *Pathogens* **2023**, *12*, 286. [CrossRef]
52. Palombo, M.; Bovo, F.; Amadesi, S.; Gaibani, P. Synergistic Activity of Cefiderocol in Combination with Piperacillin-Tazobactam, Fosfomycin, Ampicillin-Sulbactam, Imipenem-Relebactam and Ceftazidime-Avibactam against Carbapenem-Resistant Gram-Negative Bacteria. *Antibiotics* **2023**, *12*, 858. [CrossRef]
53. Mohd Sazlly Lim, S.; Heffernan, A.; Naicker, S.; Wallis, S.; Roberts, J.A.; Sime, F.B. Evaluation of Fosfomycin-Sulbactam Combination Therapy against Carbapenem-Resistant *Acinetobacter baumannii* Isolates in a Hollow-Fibre Infection Model. *Antibiotics* **2022**, *11*, 1578. [CrossRef] [PubMed]
54. Koomanachai, P.; Crandon, J.L.; Kuti, J.L.; Nicolau, D.P. Comparative pharmacodynamics for intravenous antibiotics against Gram-negative bacteria in Europe between 2002 and 2006, a report from the OPTAMA program. *Int. J. Antimicrob. Agents.* **2009**, *33*, 348–353. [CrossRef] [PubMed]
55. Bavaro, D.F.; Belati, A.; Diella, L.; Stufano, M.; Romanelli, F.; Scalone, L.; Stolfa, S.; Ronga, L.; Maurmọ, L.; Dell'aera, M.; et al. Cefiderocol-Based Combination Therapy for "Difficult-to-Treat" Gram-Negative Severe Infections: Real-Life Case Series and Future Perspectives. *Antibiotics* **2021**, *10*, 652. [CrossRef]
56. Holger, D.J.; Kunz Coyne, A.J.; Zhao, J.J.; Sandhu, A.; Salimnia, H.; Rybak, M.J. Novel Combination Therapy for Extensively Drug-Resistant *Acinetobacter baumannii* Necrotizing Pneumonia Complicated by Empyema: A Case Report. *Open Forum Infect. Dis.* **2022**, *9*, ofac092. [CrossRef] [PubMed]
57. Lopez-Novoa, J.M.; Quiros, Y.; Vicente, L.; Morales, A.I.; Lopez-Hernandez, F.J. New insights into the mechanism of aminoglycoside nephrotoxicity: An integrative point of view. *Kidney Int.* **2011**, *79*, 33–45. [CrossRef] [PubMed]
58. Najmeddin, F.; Shahrami, B.; Azadbakht, S.; Dianatkhah, M.; Rouini, M.R.; Najafi, A.; Ahmadi, A.; Sharifnia, H.; Mojtahedzadeh, M. Evaluation of Epithelial Lining Fluid Concentration of Amikacin in Critically Ill Patients With Ventilator-Associated Pneumonia. *J. Intensive Care Med.* **2020**, *35*, 400–404. [CrossRef] [PubMed]
59. Mohammadi, M.; Khayat, H.; Sayehmiri, K.; Soroush, S.; Sayehmiri, F.; Delfani, S.; Bogdanovic, L.; Taherikalani, M. Synergistic Effect of Colistin and Rifampin Against Multidrug Resistant *Acinetobacter baumannii*: A Systematic Review and Meta-Analysis. *Open Microbiol. J.* **2017**, *11*, 63–71. [CrossRef] [PubMed]
60. Tripodi, M.F.; Durante-Mangoni, E.; Fortunato, R.; Utili, R.; Zarrilli, R. Comparative activities of colistin, rifampicin, imipenem and sulbactam/ampicillin alone or in combination against epidemic multidrug-resistant *Acinetobacter baumannii* isolates producing OXA-58 carbapenemases. *Int. J. Antimicrob. Agents.* **2007**, *30*, 537–540. [CrossRef] [PubMed]
61. Chen, J.; Yang, Y.; Xiang, K.; Li, D.; Liu, H. Combined Rifampin and Sulbactam Therapy for Multidrug-Resistant *Acinetobacter baumannii* Ventilator-Associated Pneumonia in Pediatric Patients. *J. Anesth. Perioper. Med.* **2018**, *5*, 176–185. [CrossRef]
62. Trebosc, V.; Kemmer, C.; Lociuro, S.; Gitzinger, M.; Dale, G.E. Rifabutin for infusion (BV100) for the treatment of severe carbapenem-resistant *Acinetobacter baumannii* infections. *Drug Discov. Today.* **2021**, *26*, 2099–2104. [CrossRef]
63. Falagas, M.E.; Vardakas, K.Z.; Roussos, N.S. Trimethoprim/sulfamethoxazole for *Acinetobacter* spp.: A review of current microbiological and clinical evidence. *Int. J. Antimicrob. Agents* **2015**, *46*, 231–241. [CrossRef] [PubMed]
64. Raz-Pasteur, A.; Liron, Y.; Amir-Ronen, R.; Abdelgani, S.; Ohanyan, A.; Geffen, Y.; Paul, M. Trimethoprim-sulfamethoxazole vs. colistin or ampicillin-sulbactam for the treatment of carbapenem-resistant *Acinetobacter baumannii*: A retrospective matched cohort study. *J. Glob. Antimicrob. Resist.* **2019**, *17*, 168–172. [CrossRef] [PubMed]
65. Russo, A.; Gullì, S.P.; D'Avino, A.; Borrazzo, C.; Carannante, N.; Dezza, F.C.; Covino, S.; Polistina, G.; Fiorentino, G.; Trecarichi, E.M.; et al. Intravenous fosfomycin for treatment of severe infections caused by carbapenem-resistant *Acinetobacter baumannii*: A multicenter clinical experience. *Int. J. Antimicrob. Agents* **2024**, 107190. [CrossRef] [PubMed]
66. Sader, H.S.; Carvalhaes, C.G.; Streit, J.M.; Castanheira, M.; Flamm, R.K. Antimicrobial activity of cefoperazone-sulbactam tested against Gram-Negative organisms from Europe, Asia-Pacific, and Latin America. *Int. J. Infect. Dis.* **2020**, *91*, 32–37. [CrossRef] [PubMed]
67. Tuon, F.F.; Yamada, C.H.; de Andrade, A.P.; Arend, L.N.V.S.; Dos Santos Oliveira, D.; Telles, J.P. Oral doxycycline to carbapenem-resistant *Acinetobacter baumannii* infection as a polymyxin-sparing strategy: Results from a retrospective cohort. *Braz. J. Microbiol.* **2023**, *54*, 1795–1802. [CrossRef] [PubMed] [PubMed Central]

MDPI AG
Grosspeteranlage 5
4052 Basel
Switzerland
Tel.: +41 61 683 77 34

Antibiotics Editorial Office
E-mail: antibiotics@mdpi.com
www.mdpi.com/journal/antibiotics

www.ingramcontent.com/pod-product-compliance
Lightning Source LLC
Chambersburg PA
CBHW062050220926
51850CB00010B/2721